Globalization of Water
Governance in South Asia

T0298738

This cluster of books presents innovative and nuanced knowledge on water resources, based on detailed case studies from South Asia–India, Bangladesh, Bhutan, Nepal, Pakistan, and Sri Lanka. In providing comprehensive analyses of the existing economic, demographic and ideological contexts in which water policies are framed and implemented, the volumes argue for alternative, informed and integrated approaches towards efficient management and equitable distribution of water. These also explore the globalization of water governance in the region, particularly in relation to new paradigms of neoliberalism, civil society participation, integrated water resource management (IWRM), public–private partnerships, privatization, and gender mainstreaming.

Water Resources Policies in South Asia
Editors: Anjal Prakash, Sreoshi Singh, Chanda Gurung Goodrich and S. Janakarajan
ISBN 978-0-415-81198-9

Globalization of Water Governance in South Asia
Editors: Vishal Narain, Chanda Gurung Goodrich, Jayati Chourey and Anjal Prakash
ISBN 978-0-415-71066-4

Informing Water Policies in South Asia
Editors: Anjal Prakash, Chanda Gurung Goodrich and Sreoshi Singh
ISBN 978-0-415-71059-6

Water Governance and Civil Society Responses in South Asia
Editors: N. C. Narayanan, S. Parasuraman and Rajindra Ariyabandu
ISBN 978-0-415-71061-9

Globalization of Water Governance in South Asia

Editors

Vishal Narain
Chanda Gurung Goodrich
Jayati Chourey
Anjal Prakash

LONDON AND NEW YORK

First published 2014 by Routledge

2 Park Square, Milton Park, Abingdon, Oxfordshire OX14 4RN
52 Vanderbilt Avenue, New York, NY 10017

Routledge is an imprint of the Taylor & Francis Group, an informa business

First issued in paperback 2019

Typeset by
Eleven Arts
Keshav Puram
Delhi 110 035

British Library Cataloguing-in-Publication Data
A catalogue record of this book is available from the British Library

ISBN 978-0-415-71066-4 (hbk)
ISBN 978-0-367-25317-2 (pbk)

Contents

Part II: State, Markets and Civil Society: Changing Configurations in Water Management

Part III: Urbanization and Water: Emerging Conflicts, Responses and Challenges for Governance

List of Tables

List of Figures

List of Maps

List of Abbreviations

ADB	Asian Development Bank
AGM	Additional General Manager
AMC	Ahmedabad Municipal Corporation
BIWTA	Bangladesh Inland Water Transport Authority
BOD	Biochemical Oxygen Demand
BOT	Build, Operate, and Transfer
BWDB	Bangladesh Water Development Board
CBINRM	Community Based Integrated Natural Resource Management
CBOs	Community Based Organizations
CBRs	Conduct of Business Regulations
CCCIM	Central Co-ordination Committee in Irrigation Management
CDO	Chief District Officer
CFUG	Community Forest User Group
CRC	Convention on the Rights of the Child
CWRM	Comprehensive Water Resources Management
CWSSP	Community Water Supply and Sanitation Project
DAC	District Agricultural Committees
DANIDA	Danish International Developmental Agency
DAO	District Administration Office
DC	Distributary Canal
DCC	Dhaka City Corporation
DCC	District Co-ordinating Committee
DDC	District Development Committee
DEC	District Environmental Committees
DLRS	Directorate of Land Records and Survey
DoE	Department of Environment
DOH	Department of Health
DvCC	Divisional Co-ordinating Committee
DWRC	District Water Resources Committee
ETP	Effluent Treatment Plant
FCGs	Field Canal Groups
FOs	Farmers Organizations

FUG	Forest User Group
GBWSSB	Greater Bangalore Water Supply and Sewerage Project
GDP	Gross Domestic Product
GNH	Gross National Happiness
GNP	Gross National Product
GR	Government Resolution
GTZ	German Technical Assistance
GWP	Global Water Partnership
HP	High Power Committee
IDRC	International Development Research Center
IFIs	International Financial Institutions
IMD	Irrigation Management Division
INMAS	Integrated Management of Irrigation Agricultural Settlements
IPS	Intermediate Pumping Station
IRA	Independent Regulatory Authority
ISWP	Integrated State Water Plan
ITDG	Intermediate Technology Development Group
IWMI	International Water Management Institute
IWRM	Integrated Water Resources Management
JBIC	Japan Bank for International Co-operation
JE	Junior Engineer
JVP	Janatha Vimukthi Peramuna
KCWSAEIP	Kandy City Water Supply Augmentation and Environmental Improvement Project
KMC	Kandy Municipal Corporation
KNNL	Karnataka Neerawari Nigam Limited
KUKL	Kathmandu Upatyaka Khanepani Limited
KUWSDB	Karnataka Urban Water Supply and Drainage Board
KUWSSC	Karnataka Urban Water Supply and Sanitation Council
KWRA	Karnataka Water Resources Authority
LA	Local Authorities
LIG	Low Income Group
MARI	Modern Architects of Rural India
MASL	Mahaweli Authority of Sri Lanka
MC	Management Committee
MDGs	Millennium Development Goals

MGDP	Mahaweli Ganga Development Project
MKVDC	Maharashtra Krishna Valley Development Corporation
MLD	Million Liters Per Day
MoEF	Ministry of Environment and Forests
MSPs	Multi-Stakeholders' Processes
MWRRA	Maharashtra Water Resources Regulatory Authority
MWSD	Ministry of Water Supply and Drainage
MWSS	Mini Water Supply Schemes
NCWC	National Commission for Women and Children
ND-BOT	Nira Deoghar Project on Build Operate and Transfer
NEC	National Environment Commission
NGO	Non Governmental Organization
NGOFUS	Non Governmental Organizations Forum for Urban Sanitation
NSSC	Neighborhood Society Service Center
NWAB	National Women's Association of Bhutan
NWP	National Water Plan
NWSC	Nepal Water Suply Corporation
NWSDB	National Water Supply and Drainage Board
NWSSSC	National Water Supply and Sanitation Steering Committee
O&M	Operation and Maintenance
OBC	Other Backward Caste
PCC	Provincial Cooordination Committee
PIM	Participatory Irrigation Management
PMC	Project Management Committee
PO	Partner Organization
PPP	Public–Private Partnership
PUP	Public–Public Partnership
RAJUK	Rajdhani Unnayan Kartripakkha
RSCs	Regional Support Centers
RSPN	Royal Society for Protection of Nature
RTI	Right to Information
RWH	Rainwater Harvesting
RWS&S	Rural Water Supply and Sanitation
SaciWATERs	South Asia Consortium for Interdisciplinary Water Resources Studies

SEI	Stockholm Environment Institute
SLR	Sri Lankan Rupees
SSCWSS	Small Scale Community Water Supply System
STWSSS	Small Town Water Supply and Sanitation Scheme
TAP	Transparency, Accountability, and Participation
TDF	Town Development Fund
TNC	Transnational Corporation
TWSSP	Third Water Supply and Sanitation Project
ULBs	Urban Local Bodies
UNDP	United Nations Development Programme
UPWMRC	Uttar Pradesh Water Management Regulatory Commission
VDC	Village Development Committee
VWSC	Village Water & Sanitation Committee
WB	World Bank
WRC	Water Resources Council
WRDO	Water Resources Development Organisation
WRS	Water Resources Secretariat
WRS	Water Resources Strategy
WSSD	World Summit on Sustainable Development
WUAs	Water Users Associations
YDF	Youth Development Fund

Foreword

This book is part of the "crossing boundaries" project, an effort by the South Asian Consortium for Interdisciplinary Water Resources Studies (SaciWATERs) to contribute to the on-going paradigm shift in water resources management in South Asia. The "wicked problem" of water, which does not allow an easy definition let alone subsequent trouble-free, linear resolution, has also been described as an enigmatic messy problem in its and/and nature as opposed to either/or of most other substances of importance to human societies (Moss 2009). It is solid and liquid and gas; life-giving and death-delivering; a natural good and a social good and an economic good, often all of that at the same time and place, while being a cause of both co-operation and strife. This complex nature means that water issues cannot be solved by the binary either/or rock logic of bureaucratic proceduralism alone but demand more flexible fuzzy logic accommodating contradictory certitudes.

Recent decades have seen several shifts in the previously unchallenged "hydraulic missions" of government agencies (or hydrocracies) tasked with harnessing rivers and "developing" water resources within their boundaries. First, the gap between government and governance has widened with actors other than state agency hydrocrats, such as environmentalists and social activists, demanding that they be both heard and responded to. These actors are bringing into the discourse issues important to holistic and healthy water management that had been conveniently filtered out or swept under the carpet by single-mission, construction-focused hydrocracies. Second, governments are discovering that they are neither capable of mobilizing the massive capital required for water projects nor, if they do manage the capital, does their procedural fetishism leave much room for high efficiency demanded by the inexorable logic of such capital. This has opened the space for the private sector's entry into the water realm and the political acceptability of "public–private partnership." Third, the increasingly significant role played by multilateral development agencies, transnational corporations and non-government organizations has served to challenge the hegemonic sway of hydrocracies.

Until quite recently, water management was seen as the exclusive domain of technical experts working under the auspices of the state. Currently, however, participatory management with multiple stakeholders has gained increasing importance. Indeed, a World Commission on Dams was constituted (in 1998) by major players in the field to see what common ground could be found in the controversy surround the building of large dams around the world. Ten years after the Commission presented its report, the issues of controversy have not gone away, but the understanding of their complexities has been deepened (Moore et al. 2010). The notion of government as the only decision-making authority has been replaced by multi-scale, polycentric governance: it recognizes a large number of stakeholders coming from different styles of organizing with varied perceptions of what the problem itself is, what the risks are and who should bear them, as well as what solutions might be acceptable. Thus, collaborative governance with constructive engagement between divergent views is considered to be more appropriate for integrated and adaptive management regimes needed to cope with the complexity of social-ecological systems (Gyawali 2009).

At many levels, these trends have been interpreted as eroding the dominance of hydrocracies in their monochrome and unchallenged policy space. It has constricted the room for government agencies to maneuver in several areas of public policy formulation that impinge on water management. The WEHAB agenda and the urge to accomplish the MDGs have led to the creation, at the highest level of global governance, of new partnerships across the countries involving a wide range of different actors.[1] The spread of democracy across the globe, on the one hand, and the growth of private enterprise, on the other, both pose new challenges for governance processes. In the water sector, the global environmental crisis, growing poverty in urban and rural areas, continued gender inequalities, and transboundary impasse in collaboration (among other issues) all point to the need for a different governance approach to water use and management. Multi-stakeholder platforms, wherever they have been established and practiced, while not providing magic bullet solutions, have, however, helped in the process of constructive engagement–"negotiations" in short– by providing the forum where the views of the "other side" can both be heard and responded to (Dore et al. 2010). As a result of such developments, water (and other natural resources) management has been undergoing major paradigm shifts that may be the harbinger of more healthy and less conflict-ridden developments in the future.

All these developments and challenges have led academicians and practitioners all over the world to express increasing concerns over the inadequacy of current measures of universal economic performance, in particular those based on GDP figures. The concern is about the relevance of these indicators as measures of societal and human well-being and equity as well as measures of economic, environmental, and social sustainability. Current well-being has to do with both economic resources, such as income, and with non-economic aspects of peoples' life–what they can do, what they should do, and how they feel and value the natural environment they live in. Whether these levels of well-being can be sustained over time depends on whether stocks of capital that matter for our lives (natural, physical, human, social) are passed on to future generations in as good a shape, if not better, than how we ourselves have inherited them.[2] The big question is–will South Asia make seminal contributions to this debate through indigenous approaches or will it simply follow and parrot concepts developed elsewhere?

Bhutan, where a conference was organized by SaciWATERs in May 2010, which has led to the chapters in this book, stands out by its introducing the concept of "Gross National Happiness" (GNH) into the global discourse. It may have many operational difficulties, but it certainly is a paradigm challenge that has to be engaged with constructively, especially by those in South Asia. As far as water is concerned, South Asia's civic movements are rich in new explorations, which have crossed many boundaries, not just the political and administrative but also conceptual and disciplinary. They have allowed us to see the "social construction" of water by different social solidarities or stakeholders, the implications they have for the choice of variegated water technologies ranging from age-old traditional to modern industrial, and the plural perceptions of risk and equity that have guided their invention as well as use and adaptation.

The Arun and Mahakali debates in Nepal, those of Tehri and Narmada in India, Eppawala in Sri Lanka, flood action plan in Bangladesh or the Kalabagh in Pakistan have all deepened our understanding of the perils of the unbridled hydraulic mission paradigm and the reflexive discourse on modernity demanded by our times (Gyawali et al. 2006). And it is not that there are no success stories in our neighborhood to build on. Alternative thinking on water management in Calcutta has shown how urban wastewater can be cleaned and the nutrients harvested through algae and fisheries before they are lost to the sea. Water harvesting and restoration of traditional tanks in the dry parts of western India have been inspirational measures. Those who worked to restore ponds in the mid-hills of the Nepal Himalayas found, to their surprise, that these measures also prevented landslides from growing, increased maize production with the increase in soil moisture, and gave extra weeks of life to mountain springs, thus reducing water stress for people and livestock alike.

This book is both a marker and a continuing journey by the collegiums of South Asia and beyond engaged with SaciWATERs. The Thimpu conference in May 2010 focused on the implications of the trends and shifts in water management of this region under the globalization process and asked the question: how are these trends changing water management practices, ownership and access? How are water management policies being reformulated? What roles have international and multilateral institutions played in influencing the direction and content of water sector reform processes? What room do governments have to maneuver vis-a-vis the international political and economic order in the management of their reform processes? How is the growth of private enterprise influencing access to water? And finally, what are the implications of these trends for human and societal well-being? The workshop was attended by about 50 researchers, academicians, practitioners and students from South Asian countries. Of the 25 papers presented at the conference, seventeen appear in this volume following a rigorous peer review process and the incorporation of the debates at the workshop.

I am confident that this book will provide a new searchlight on the path ahead and contribute to the public debate around the democratization of water governance in South Asia.

Dipak Gyawali
Academician, Nepal Academy of Science and Technology
Former Minister of Water Resources
Kathmandu

Notes

1. UN Secretary-General Kofi Annan proposed at the Johannesburg World Summit on Sustainable Development (WSSD) in 2002 five key areas for particular focus—Water, Energy, Health, Agriculture and Biodiversity (WEHAB).Available at http://esl.jrc.ec.europa.eu/dc/wehab/WEHAB_Indicators.htm (accessed May 2, 2013).
2. Report by the Commission on the Measurement of Economic Performance and Social Progress. Available at http://www.stiglitz-sen-fitoussi.fr/en/index.htm (accessed May 2, 2013).

References

Dore, J., J. Robinson and M. Smith. 2010. "Negotiate: Reaching Agreements over Water," *Gland: IUCN, Water and Nature Initiative.* Available at http://data.iucn.org/dbtw-wpd/edocs/2010-006.pdf (accessed May 2, 2013).

Gyawali, D. 2009. "Pluralized Water Policy Terrain = Sustainability and Integration," *SAWAS*, 1(2): 193–199. Available at http://www.sawasjournal.org/templates/sawas/images/PluralizedWaterPolicyTerrain.pdf (accessed May 2, 2013).

Gyawali, D., Allan, J.A. et al. 2006. "EU–INCO Water Research from FP4 to FP6 (1994-2006): A Critical Review." Luxembourg: Office for Official Publications of the European Communities. Available at http://ec.europa.eu/research/water-initiative/pdf/incowater_fp4fp6_rapport_technique_en.pdf (accessed May 2, 2013).

Moore, D., J. Dore and D. Gyawali. 2010. "The World Commission on Dams + 10: Revisiting the Large Dam Controversy," *Water Alternatives*, 3(2): 3–13. Available at http://www.water-alternatives.org/index.php?option=com_content&task=view&id=139&Itemid=1 (accessed May 2, 2013).

Moss, J. 2009. "Water Ethics and Business," in M.R, Llamas, L. Martinez-Cortina and A. Mukherji (eds) *Water Ethics.* London: CRC Press, Taylor and Francis Group.

Acknowledgements

Ideas for this book evolved through discussions and several rounds of reflection among the editors on the implications of the globalization of governance for water management in South Asia. After the editors met at a regional workshop in May 2009, in Kathmandu, the idea of organizing a workshop on this theme in Bhutan took shape. The editors profoundly acknowledge the support of the Royal Society for the Protection of Nature (RSPN), Thimphu, Bhutan, for hosting the workshop. The papers presented at the workshop formed the base of the contributions in this volume. In particular, we would like to acknowledge Dr Lam Dorji, Executive Director of RSPN, Ms Tshering Lhamtshok, Mr Ugyen Lhendup, Mr Phuntsho Yonten, Mr Pema Gyamtsho, Mr Sonam Jamtsho, Ms Yangchen Lhamo, Mr Chenning Dorji, and Mr Cheten Dorji–who did all the arrangements for holding the research workshop.

Financial grant from the Government of the Netherlands, through the Crossing Boundaries Project, for the workshop and for supporting the travel and stay of the contributors and participants is gratefully acknowledged. The editors also thank the office staff at SaciWATERs–particularly Hemalatha Paul, Office Manager, and C. Sreenivasulu, Finance Manager–for the generous support provided at various stages of the book production process. Finally, the editors acknowledge the comments and suggestions made by the participants of the workshop that helped crystallize ideas around the theme.

Introduction

The Globalization of Governance: Transforming Water Management in South Asia?

Vishal Narain, Chanda Gurung Goodrich, Jayati Chourey and *Anjal Prakash*

The expression "globalization of governance" is open to diverse meanings as it is composed of two terms, which themselves have several connotations. P. G. Cerny, G. Menz and S. Soederberg (2005: 9) rightly note that "there are almost as many definitions as there are scholars and actors writing and thinking about globalization." J. B. Gelinas (2003), noting the many ways in which the term can be interpreted, sees globalization as a system, a process, an ideology and an alibi. As a system, it represents the total control of the world by supranational economic interests. As a process, it represents a series of actions carried out in order to achieve a particular result. As an ideology, it represents a coherent set of beliefs, views and ideas determining the nature of truth in a given society. Its role is to justify the established political and economic system and make people accept it as the only one that is legitimate, respectable and possible. As an alibi, globalization is presented as a natural, inevitable and irresistible phenomenon that lets the major economic and political decision-makers off the hook.

G. Bertucci and A. Alberti (2003: 17) define globalization as "the increasing flows between countries of goods, services, capital, ideas, information and people that produce cross-border integration of economic, social and cultural activities;" "the closer integration of the countries and peoples of the world which has been brought about by the enormous reduction of costs of transportation and communication, and the breaking down of artificial barriers to the flows of goods, services, capital, knowledge and (to a lesser extent) people across borders" (Stiglitz 2002: 9). What is implicit in these conceptualizations is the weakening (or blurring) of political boundaries and the greater flows of ideas and influences across the globe.

Governance refers to the exercise of control and authority in the use and allocation of a nation's resources.[1] The State, markets and civil society are recognized as the three main actors in governance. Recent ideas of governance point to changing relationships among these, most notably the increasing role of actors other than the State. Terms, such as "polycentric governance" or "network governance," are used to capture these transformations (Mathur 2009; Pierre 2000).

We could then define the globalization of governance as a process in which control over policy-making, at the level of nation-states, has shifted beyond their political boundaries, and in which actors outside the state (both locally and globally) have come to play a greater role. Though this process has been a hallmark of the 20th century, it has taken a more specific form over the last three to four decades. This trend has been shaped by factors, such as (*a*) the growing role of multilateral institutions like the World Bank, the World Trade Organization (WTO) and the International Monetary Fund (IMF) that are able to influence policy-making in many developing countries; (*b*) the rise of transnational companies and NGOs that often challenge state authority; (*c*) the importance of international summits and conventions that compel governments to act in specific ways; (*d*) the global spread of democracy that seems to bring some convergence in governance mechanisms across the globe; and (*e*) the spread of information and communication technology, which makes the world increasingly borderless.

Trends Shaping the Globalization of Governance

Perhaps, the most important trend shaping the globalization of governance—and one with clear implications for water management and water sector reforms in South Asia, as demonstrated in several chapters in this volume—has been the increasing role of multilateral and bilateral institutions. This has implications for not only national level governance but also local livelihoods and vulnerability; the impacts of these institutions as well as the policies and laws that they urge governments to formulate and implement, trickle down from global to local levels.

A case in point is the neoliberal agenda in the form of a heightened role of markets as espoused by the World Bank in the 1980s. The role of the free market was built on a narrative of an "inefficient state." This took the form of conditions attached to financial support programs, which (many argue) shifted the locus of control over decision-making from the national to the international arena. The new economic policies, which India and many other developing countries embraced in the 1990s, for instance, could be seen to be located in these neoliberal discourses promoted by such international institutions.

The growing role of multilateral institutions in influencing public policy (in developing countries) has led to wide-ranging debates on the autonomy of the State; while some scholars like H. Chang (2006) present a gloomy picture, contending that this has severely eroded State autonomy, others like Shalini Randeria (2003) have argued that the State only selectively implements what it seeks to, giving rise to the notion of the "cunning state."

Likewise, international corporations, too, have grown in size, number and significance and pose a challenge to State authority. At the beginning of the 1980s, the best-funded and best-positioned corporations acquired the status of transnational corporations (TNCs) (Gelinas 2003). Many of these mega enterprises are known to stand beyond or above the nation-state. With increasingly numerous mergers, takeovers and alliances, the concentration of wealth, at the top, reached such a degree that economic power seemingly succeeded in liberating itself from the national legal framework and from governmental control. The information revolution of the 1980s—marked by an extraordinary influx of new telecommunications, computerization and automation—further played an important role in transforming these corporations into TNCs. These tools allowed them to go beyond the former limits of time, space, borders, languages, and cultures.

These TNCs share a number of features in common, such as (*a*) a huge direct foreign investment capacity, embodied in a network of subsidiaries and sub-contractors across the world; (*b*) a financial strategic potential for carrying out mergers and forging alliances capable of concentrating supply and demand to neutralize and, ideally, eliminate competition; and (*c*) a capacity to relocate and delocate production units anywhere in the world where labor is cheaper and where environmental and social laws and regulations are least restrictive.

In a similar vein, transnational non-governmental organizations (NGOs) have risen in significance and have come to influence governance processes globally. In particular, NGOs have begun to play an important role in challenging authoritarian rule and pushing donor governments towards a more proactive role against repressive regimes (Schmitz 2006). Four decades ago, about 1,000 NGOs operated mostly at the local level; at the turn of the century, the United Nations reports that almost 30,000 NGOs operate internationally. Moreover, NGOs are invited to participate in global meetings, such as the World Summit on Sustainable Development (WSSD) 2002, which can be taken as a positive indication of the growing recognition of their role and position. Needless to say, the changing configurations between State, civil society and private corporations have raised new issues about participation, voice and accountability in governance processes (Blair 2000; Paul 1992).

The "Good Governance" Paradigm

Associated closely with the globalization of governance is the concept of "good governance." It is used widely by bilateral and multilateral organizations, NGOs, academics, and researchers and is a recurring theme in national and international discourses and debates, providing a basis for directing financial resources in specific directions. This concept has had a visible presence in the water sector in South Asia, with many policy changes in water management seemingly deriving their justification from it, as several chapters in this book would demonstrate.

The "good governance" paradigm was floated and advocated by the World Bank in the 1990s in response to the unsatisfactory results in some of the reform programs supported by it.[2] In this context, improving systems of governance came to be identified as a means of bringing about sustainable economic development. It was argued that for better economic management, fundamental changes were needed within the political system along with appropriate institutional reforms, and mere injection of financial resources was far from being enough. In this context, the concept of "good governance" was propagated as the means to secure more effective utilization of financial resources that were being introduced by the World Bank into the economies of the developing world.

The World Bank's vision of "good governance" envisages a bureaucracy imbued with professional ethos, an accountable executive arm of the government, a strong civil society participating in public affairs, and citizens behaving under the rule of law (Shylendra 2004). Lately, this concept has been broadened by official aid agencies incorporating political dimensions to include and emphasize participatory development, democratic systems that promote transparent and accountable societies, and civil society participation. In essence, the concept has been widened to include notions of civil society participation in public affairs.

The Prescriptive and Descriptive Dimensions of "Good Governance"

H. S. Shylendra (2004) notes the usage of this expression both in a prescriptive and a descriptive sense. In a descriptive sense, this phenomenon is seen as emerging due to a changing role of the State, in developed as well as in developing countries. This is happening in a context of neoliberal ideologies, globalization and liberalization. The State is cutting back expenditure while other actors are acquiring a pre-eminent role, leading to what is called a "pluralization of the State."

In a prescriptive sense, the "good governance" paradigm refers to enabling the right manner of exercise of power for attaining better economic development in a country as well as simultaneously enhancing the scope for participation and democratic processes (World Bank 1994).

In this sense, it could be seen as comprising a wish list of elements that are essential for improving governance systems. Four important ideas are thus emphasized in the thrust on "good governance"–improving public sector management, ensuring accountability of public and private sectors, creating an appropriate legal framework for development, and promoting transparency and information. However, the concept has also evoked a number of criticisms; namely, that it is too utopian, excessively prescriptive and that multilateral institutions propagate it in a "holier than thou" position.

Globalization, Urbanization and the Growth of Urban Agglomerations

In the wake of globalization, there has been considerable interest among urban theorists and scholars in the emergence of urban agglomerations, particularly in the relative roles of global and local forces in shaping urbanization processes as well as the growth of large urban agglomerations. The opening up of the economies of the South Asian countries, since the 1990s, has brought in many changes in the large cities of this region, particularly the Indian cities. This is a process in which both local and global agencies are understood to have played a role, though their relative influence has varied.

Annapurna Shaw and M. K. Satish (2007) make an effort to separate the local influences (shaping urbanization) from the global influences, and argue that there has been a general tendency, both in the popular press as well as mainstream academics, to connect all forms of change in the economy, society and culture to globalization; they maintain that doing so has obfuscated rather than illuminated the actual processes of change underway and their implications. In a similar vein, G. Shatkin (2007) points to the failure to explore local agency, in relation to globalization and urbanization, describing the emergence of modern cities as being shaped by the negotiated environment in which local and global actors together influence the built environment.

Within the South Asian region, the growth of cities has been led by a mix of factors; on the one hand, neoliberal policies have given greater space to large transnational corporations of the kind described earlier; governments have also introduced policies for a greater involvement of private corporations. At the same time, local networks and alliances, at various levels, have given a boost to these processes.

A very interesting example of this is provided in the analysis of Mumbai's urban development by Liza Weinstein (2008). She describes how this development has been supported by the local mafia criminals who have worked closely with the State–that is, with politicians and bureaucrats–within a context of globalization. Local criminal syndicates (often with global connections) have seized political opportunities created by these shifts to gain influence over land development. The rise of Mumbai's organized criminal activity, in the 1950s, was closely linked to India's macro-economic policies with strict regulation of imports fuelling the growth of black market smuggling; liberalization and deregulation, since the early 1990s, have diminished the demand for smuggled consumer goods, and criminal syndicates have since diversified their operations. With skyrocketing real estate prices, bolstered by global land speculation, the mafia began investing in property development; supported by an illegal nexus of politicians, bureaucrats and the poor, the mafia has emerged as a central figure in Mumbai's land development politics.

Notwithstanding the relative roles of local and global forces in shaping the processes of urbanization, a common feature of this growth of South Asian cities has been a predominant role of the real estate and construction sectors, which has altered the face of these cities. It has influenced the pace of urbanization and altered the demand for water along with patterns of water allocation across rural and urban uses. Considerable land use change has occurred in the peripheries of these large cities, with wide implications for the use of natural resources like water.

Transforming Water Management in South Asia?

The chapters in this volume seek to examine the implications of these processes of the globalization of governance for water management in the region, particularly their influence on human well-being, the functioning of water management utilities and the relationship between State, markets and civil society. They are a selection of papers presented at the 4th Regional Research workshop on Globalization of Governance; Implications for Water Management in South Asia, organized by the South Asian Consortium for Interdisciplinary Water Resources Studies (SaciWATERs) in collaboration with the

Royal Society for the Protection of Nature (RSPN), in Thimphu, Bhutan, from May 3–5, 2010.

Neo-liberal reforms and global discourses on governance, of the kind described in the preceding sections, have pervaded all five countries in South Asia represented in this volume–India, Sri Lanka, Bangladesh, Nepal, and even land-locked Bhutan. Integrated water resources management, full cost recovery, tariff rationalization, and multiple stakeholder participation have been the new mantras for water management in the region. These are built around the narratives of the "inefficient state," on the one hand, and discourse of the "free market," on the other, propagated and echoed by donors and funders as well as by global coalitions and networks. What have been the implications of these reforms for human well-being and improved access to water? Have neoliberal reforms improved the access of the poor and underprivileged to water? Or are there inherent contradictions in the content of these discourses that militate against equity and improved access for all?

Part 1 of the volume assesses the empirical experience on this front. Seema Kulkarni provides a context for this discussion by providing a review of the various conceptual connotations of human well-being and how the globalization of governance has changed the discourses around water management from predominantly technical to social, economic and institutional. She describes the changing nature of discourses around water management in the region as well as the shifting narratives around equity, justice and gender empowerment. She goes on to argue that these new discourses have, however, changed little by way of improving the access of women to water, for whom it has been difficult to overcome social constraints and norms. This view is reinforced by Deepa Joshi's analysis, which notes that the "rhetoric of reform" seems to have been just a "rhetoric" that has failed to impact the perverse niche created by the intersection of caste, class and gender in which poor Dalit women find themselves entrenched. Neoliberal approaches have indeed shown a very weak understanding of the caste, gender and water nexus.

On the contrary, as shown in the chapter by N. I. Wickremasinghe, drawing upon a study in Sri Lanka, neoliberal principles enshrined in such stipulations that make everyone pay for a water connection, means that many who cannot pay are by definition left out. These are some of the weaknesses of demand-driven approaches from a standpoint of ensuring equity in access to water—an idea that seems to echo in several chapters in the book.

Gongsar Karma Chhopel's chapter on Bhutan shows how the global discourses, particularly those of IWRM and gender, have been embraced in the country and how they have been woven in the country's traditional and religious beliefs, under the unique Gross National Happiness index. However, "gender" here too seems to have been "rhetoric" as most of the projects and programs address typical gender issues, such as domestic violence, training on business skills, among others, while the crucial issues of equity, empowerment and altering gender relations remain unaddressed.

Part II of the book traces the wider implications of these processes for how water is managed in the region, and how they have impacted the relationship between the State, markets and civil society as major actors in governance. Chapters in this part note in particular the predominant role of bilateral and multilateral institutions in influencing the direction and content of water reforms. They also engage with the experience with new and emerging forms of water institutions, such as independent regulatory authorities, multiple stakeholders' platforms, privatization, and public–private partnerships as well as the relationship between different reform instruments to assess their compatibility.

The analysis by Divya Badami Rao and Shrinivas Badiger points to the weaknesses in the functioning of the model of Independent Regulatory Authorities (IRAs) in the Indian state of Karnataka, seeing it as a model imported from abroad; they describe the process of setting up of independent regulatory authorities as a "clever process of depoliticization of the water sector to institutionalize new principles of water governance as full cost recovery or to prioritize economic uses of water (p. 178 of this book)." Based on their critique of the working of IRAs in Karnataka, they question whether these are indeed an appropriate answer to the ills of the sector. In a similar vein, Sachin Warghade and Subodh Wagle point to the inconsistency in several of the reform instruments in the state of Maharashtra. Similarly, Dhruba Pant, in his analysis of a "Platform Approach" to secure Integrated Water Resource Management (IWRM) in Nepal, notes that policy formulation in itself is inadequate unless it is backed by legislative and institutional reforms for its effective implementation. In other words, planning the implementation of such initiatives needs to be part of the policy formulation process itself. E. R. N. Gunawardena suggests that though governments have pursued decentralization policies under the influence of donor-funded projects, there is reluctance to devolve authority and control to local levels. In what would seem like an overall message from Part II, Gunawardena makes a rather radical suggestion–that strengthening existing systems of governance might be more relevant than supplanting them with exogenous governance models.

The analysis of a small-scale water supply and sanitation scheme in Nepal by Hari Shrestha seems to support the "small is beautiful" view, demonstrating how small-scale community-managed water supply systems fill a void left both by the State and the market. When the State–even with all the efforts to reform itself and to improve its functioning–fails to meet the requirements of water supply, it is a community-based water supply that seems to have provided a viable alternative, though only at a small scale. The water that they have been able to thus distribute amongst themselves is better in terms of its quality and much cheaper than what is provided by water tankers. This observation is echoed by Prakash Gaudel who demonstrates ways and means of the inclusion of the "Poor" and "Very Poor" households as beneficiaries of a water supply and sanitation scheme in Nepal.

Chapters in Part III of this volume look at the challenges thrown up by the processes of urbanization–shaped by a curious intersection of globalization and local agency and power as demonstrated earlier–for water management in the region. The analyses clearly show emerging conflicts between rural and urban water users and uses. At the receiving end have been the poor and the marginalized groups who lose out on access to water in terms of quality and quantity, as suggested by Shrinivas Badiger and his colleagues, and Parthasarathy and Soumini Raja. Bijaya and Sushmita Shrestha's analysis shows how urbanization processes have eroded the fit between technology, society and settlement in the Kathmandu Valley; new institutional frameworks failed to meet the requirements of water that a growing city posed. The failure of the State mechanism (to respond to the requirements of a growing city) is also echoed in the respective analyses of Sunil Thrikawala for the city of Kandy, in Sri Lanka, and for Dhaka city by Shah Jahan Mondal and his colleagues. In a pioneering chapter, Dibesh Shrestha and Ashutosh Shukla show how water tanker markets fill a void left by the growing demand supply gap in Kathmandu. They estimate both the value and volume of water transactions to be quite significant in magnitude.

Emerging Governance Regimes: Addressing the Politics of Policy and Planning

A case is often made for a better understanding of the politics of policy and the processes of policy change as well as the underlying narratives and discourses, and how they reflect the international balance of power among major drivers of policy change (Mollinga 2003; 2008). The chapters in this volume help us understand the genesis of several reform initiatives in the region, and place their underlying narratives and discourses in both their political and historical context, and contribute to a further analysis of the impacts of these reforms and discourses on the poor and marginalized, particularly women. To this effect, they make a valuable contribution to a body of work that is emerging and should grow in the years to come.[3] They lead us to make a case–as does Seema Kulkarni–for "re-imagining" or re-conceptualizing both the water sector and the communities that are seen as its beneficiaries. Neoliberal reforms in themselves do little to promote equity and improve access and need to explicitly address issues of caste, class and gender, both in the design as well as in the implementation of programs.

Another important message of these chapters (particularly in Part II) is that state governments, when contemplating reforms, need greater clarity in what they wish to accomplish through the reforms, rather than carrying them out under donor persuasion or pressure. This can often result in contradiction among different kinds of reform measures. They also suggest how the relationship between the State, markets and civil society has changed in the context of globalization. State authority has most certainly been diluted–by the role of international institutions on the one hand, and by markets and civil society institutions that have stepped in when even a "neoliberally reformed State" has failed to deliver, on the other. Authors point out alternatives to the regulatory regimes, which have sprung up in the garb of regulatory authorities, and suggest involving the poor and the marginalized in water management programs.

A common thread in all the chapters is the equity in access to water, especially the deprivation of those at the periphery of large cities, as demonstrated in Part III. In the process of urbanization and the rapid growth of cities, there continue to be pockets of disadvantage where deprivation is likely to be increased by close proximity to an affluent majority (Errington 1995). It has been widely observed that current processes of urbanization in the region have tended to leave out the poor and the marginalized, with a strong potential for conflict. "Theories and praxis of neoliberal urbanism and the enforcement of regulatory regime in cities and their regions are getting intrinsically associated with such resistances and struggles signifying a radical politics of contestation that would finally decide for who the cities and their spaces are meant for" (Banerjee-Guha 2009: 106).

To unpack these processes of deprivation and marginalization of those at the periphery of large cities requires explicit attention to the politics of urban planning. Shatkin (2007: 9) notes that much of the literature on urban planning has failed to consider the significance of "politics and power relations," assuming instead that all actors have equal amounts of power at their disposal. Different interests compete with each other to make the city what they want to be for their own survival (Roy 2004). City planners end up catering only to the needs of those whose interests are more powerful–planning, in essence, is a battle to assert dominance over areas of space which are negotiated and contested by many different actors; in this space of contestation, those at the periphery themselves have barely any role to play.

A recent World Bank study (2009) notes that trying to restrict rural–urban migration can be counterproductive because limiting density and diversity stifles innovation and productivity; rural–urban transformations are best facilitated when policy makers recognize the economic interdependence among settlements. The debate should instead be about the efficiency and inclusiveness of the processes that transform a rural economy into an urban one. In the case of water, perhaps, we need to start by understanding the various claimants over water in peri-urban areas, identify their uses and the various means adopted by them in order to meet their water needs, and then create processes for negotiation among the various stakeholders. While a separation of rights in water from rights to land–at the heart of rural–urban conflicts–is imperative, it remain difficult to operationalize (Narain 2009).

In general, as the chapters in Part III point out, policy-makers and planners need to appreciate and understand that the "rural" and "urban" are closely related, not only in terms of the flow of goods and services, which support both rural and urban livelihoods, but also through the flow of natural resources like water. The dichotomy between "rural" and the "urban" water supply, as used in conventional planning and reporting of progress, needs to be steered away from, and the flows of water between these areas need to be better understood and documented. Planning water management at a regional level will be necessary.

Acknowledgements

The authors thank Gopa Samantha and Priya Sangameswaran for their useful comments on an earlier draft of this chapter.

Notes

1. For conceptual discussions on governance, see Mathur (2009), Pierre (2000), Pierre and Peters (2000), Stoker (1998), and the World Bank (1994).
2. For a detailed discussion of the genesis of the concept, also see Mathur (2009).
3. For such studies in the international context, see Huitema and Meijerink (2009). For studies in the Indian context, see Narain (2008, 2009a), Mollinga (2003) and Urs and Whittel (2008).

References

Banerjee-Guha, S. 2009. "Neoliberalising the 'Urban': New Geographies of Power and Injustice in Indian Cities," *Economic and Political Weekly*, 44(22): 95–107.

Bertucci, G. and A. Alberti. 2003. "Globalisation and the Role of the State: Challenges and Perspectives", in D. R. Rondinelli and G. S. Cheema (eds), *Reinventing Government for the Twenty-first Century: State Capacity in a Globalizing Society*, pp. 17–31. Bloomfield, CT: Kumarian Press.

Blair, H. 2000. "Participation and Accountability at the Periphery: Democracy and Local Governance in Six Countries," *World Development*, 28(1): 21–39.

Cerny P. G., G. Menz, and S. Soederberg. 2005. "Different Roads to Globalization: Neoliberalism, the Competition State, and Politics in a More Open World," in S. Soederberg, G. Menz and P. G. Cerny (eds) *Internalizing Globalization: The Rise of Neoliberalism and the Decline of National Varieties of Capitalism*, pp. 1–30. International Political Economy Series. Basingstoke: Palgrave Macmillan.

Chang, H. 2006. "Policy Space in Historical Perspective with Special Reference to Trade and Industrial Policies," *Economic and Political Weekly*, 41(7): 627–34.

Errington, A. 1995. "The Peri-urban Fringe: Europe's Forgotten Rural Areas," *Journal of Rural Studies*, 10(4): 367–75.

Gélinas, J. B. 2003. *Juggernaut Politics: Understanding Predatory Globalization.* London; New York: Zed Books.

Mollinga, P. P. 2003. *On the Waterfront: Water Distribution, Technology and Agrarian Change in a South Indian Canal Irrigation System.* Wageningen University Water Resources Series, Vol. 5. Hyderabad: Orient Longman.

———. 2008. "The Water Resources Policy Process in India: Centralization, Polarization and New Demands on Governance," in V. Ballabh (ed.), *Governance of Water. Institutional Alternatives and Political Economy*, pp. 339–70. New Delhi: Sage Publications.

Mathur, K. 2009. *From Government to Governance: A Brief Survey of the Indian Experience.* New Delhi: National Book Trust.

Narain, V. 2008. "Crafting Institutions for Collective Action in Canal Irrigation: Can We Break the Deadlocks?," in V. Ballabh (ed.), *Governance of Water: Institutional Alternatives and Political Economy*, pp. 159–73. New Delhi: Sage Publications.

———. 2009a. "Where Does Policy Change Come From? And Where Does It End Up? Establishing Water Users' Associations in Large-Scale Irrigation Systems in India," in D. Huitema and S. Meijerink (eds), *Water Policy Entrepreneurs. A Research Companion to Water Transitions around the Globe*, pp. 120–36. Elgar Original Reference Series. Cheltenham: Edward Elgar.

———. 2009b. "Gone Land, Gone Water: Crossing Fluid Boundaries in Peri-urban Gurgaon and Faridabad, India," *South Asian Water Studies* 1(2): 143–58.

Pierre, J. 2000. "Introduction: Understanding Governance', in J. Pierre (ed.), *Debating Governance. Authority, Steering and Democracy*, pp. 1–10. New York: Oxford University Press.

Pierre, J. and B. G. Peters. 2000. *Governance, Politics and the State.* Political Analysis Series. Basingstoke: Palgrave Macmillan.

Randeria, S. 2003. "Glocalization of Law: Environmental Justice, World Bank, NGOs and the Cunning State in India," *Current Sociology*, 51(3/4): 305–328.

Paul, S. 1992. "Accountability in Public Services: Exit, Voice and Control," *World Development*, 20(7): 1047–60.

Roy, D. 2004. "From Home to Estate," *Seminar*, 533: 68–75. Also available online at http://www.india-seminar.com/2004/533/533%20dunu%20 roy.htm (accessed March 12, 2013).

Schmitz, H. P. 2006. *Transnational Mobilization and Domestic Regime Change:*

Africa in Comparative Perspective. International Political Economy Series. Basingstoke; New York: Palgrave Macmillan.

Shatkin, G. 2007. "Global Cities of the South: Emerging Perspectives in Growth and Inequality," *Cities,* 24(1): 1–15.

Shaw, A. and M. K. Satish. 2007. "Metropolitan Restructuring in Post-Liberalized India: Separating the Global and the Local," *Cities,* 24(2): 148–63.

Shylendra, H. S. 2004. "The Emerging Governance Paradigm and Its Implications for Poverty Alleviation and Equity." Working Paper 182, Institute of Rural Management, Anand, India.

Stiglitz, J. 2002. *Globalization and Its Discontents.* New Delhi: Penguin Books.

Stoker, G. 1998. "Governance as Theory: Five Propositions," *International Social Science Journal,* 50(155): 17–28.

Urs, Kshitij and R. Whittel. 2009. *Resisting Reform? Water Profits and Democracy.* New Delhi: Sage Publications.

Weinstein, L. 2008. "Mumbai's Development Mafias: Globalization, Organized Crime and Land Development," *International Journal of Urban and Regional Research,* 32(1): 22–39.

World Bank. 1994. *Governance: The World Bank's Experience.* Washington, DC: World Bank.

———. 2009. *World Development Report 2009: Spatial Disparities and Development Policy.* Washington, DC.: World Bank.

PART I

IWRM, Well-being and Gender

1

Gender, Water and Well-being

Seema Kulkarni

—

After the 1990s, the discourse in the water sector, globally, is changing. New concepts, language and terminologies are emerging while participatory institutions, legislations, new pricing policies and regulatory mechanisms are becoming the norm. The dominance of the technical aspects of water management has been replaced by that of the economic and institutional ones. As part of the various strategies proposed to counter the emerging crisis in the water sector, Integrated Water Resource Management (IWRM) is the most widely discussed and debated concept. IWRM as a terminology gained currency through the platform of Global Water Partnership which, along with the World Water Council, has been influencing the global water policy. It is broadly defined as "the co-ordinated development and management of water, land and related resources in order to maximize economic and social welfare without compromising the sustainability of ecosystems and the environment" (for more information, see GWP). It is considered as a cross-sectoral policy approach to tackle the fragmented thinking and practice in the water sector—argued as the main reason for its unsustainable use and inequitable access. IWRM, therefore, looked promising in terms of the role it would play in addressing the challenges of increasing water scarcity and its mismanagement, and increasing inequities in the sector. More than a decade has been spent in analyzing the concept of IWRM and its relevance to different regions and its implications for practice, using a large number of resources and tool kits. The major critique around IWRM has been the lack of political edge in the way it is conceptualized and implemented. Overemphasis on river basin organizations which emerged as part of the global discourse on IWRM without a corresponding analysis of the land and water linkages, the social and political contexts limits the relevance of IWRM especially so in the South Asian context (Mollinga et al. 2006).

The human rights discourse along with the feminist and ecological perspectives have been raising issues of sustainable use and equitable access to resources for all. These discourses are refraining from using the IWRM terminology as its scorecard has not necessarily been positive, especially in addressing structural inequities manifested in unequal access to water and its governance institutions for women, exploited castes and classes, ethnic minorities and indigenous groups, among others.

These new discourses have been able to challenge the dominance of the technical or hydrological disciplines in water and also highlight the fact that solutions cannot be found in the economic and institutional arena alone. At a broader level, these perspectives have been able to assert that access to, management of, and use of water is shaped not only by technical, economic and environmental factors but by social factors as well. For example, political economists as well as geographers and political ecologists have helped to focus attention on water as a strategic, increasingly scarce, and a deeply socially and culturally embedded resource around which class, gender, and globalization struggles pivot (Barlow and Clarke 2003). Human development discourses focus more on domestic water, drawing out the human rights and equity dimensions of access to safe and potable water (UNDP, 2006; Harris 2009). More recently, the hegemonic logic of water privatization is brought out through myriad examples of dispossessing the poor throughout the globe and the struggles of resistance around it (Swyngedouw 2006).

Consequently, these parallel strands of thought and action lead us to the significance of human well-being and equity discourse in the water sector. Human well-being and equity, especially gender equity, are new and inter-related challenges for water managers, which have emerged in response to failures of the past. Assessments of progress have largely been based on the overall developments in economic indicators such as increased incomes, as a result of improved water access, thereby ignoring the unequal access to water and its benefits to women and other socially differentiated groups. Also, the gender concepts of power inequality and the importance of transformation to ensure that women achieve positive and lasting changes in status are largely absent as that would entail a shift in power and implications for institutional hierarchies/hegemonies and social relations.

Part I (of this book) comprises four chapters that look at how gender, caste and class are being addressed in the water sector and the way gender and/or human well-being concerns are taken into account in water policies and practice. It also looks at the impact of water policies and programs on poor men and women of diverse social groups. The first chapter discusses broadly the theme of well-being and its application in understanding the impacts of policies and programs in the water sector, based on empirical studies in India. The chapter concludes by stressing the relevance of using the "well-being framework" in water (in its wider sense) to capture not merely access to water but also the capacity to influence decisions in water use that articulate social justice.

The second chapter by Deepa Joshi discusses the complex interplay of caste and gender in the water sector, by presenting the traditional systems, the centralized welfare state period and the present neoliberal reforms. She shows through documents and empirical research how, in all three situations, the question of both caste and gender remains completely unaddressed and concludes with a question on how and whether caste and gender will at all be addressed in the water sector. The third chapter by N. I. Wickremasinghe (from Sri Lanka) discusses the limitations of the demand-driven approach of water supply in addressing issues of equity, social justice and women's empowerment. It studies the different reasons for the exclusion of the marginalized communities in these demand-driven approaches, characterized by community contributions towards capital costs of the scheme. The author argues that these contributions become the main barrier for achieving the well-being of marginalized communities and women.

The fourth chapter by Karma G. Chhopel (from Bhutan) discusses the situation in Bhutan which, he argues, is in favor of well-being and happiness for all as the concept of Gross National Happiness (GNH)–the national indicator of the country–defines Bhutan's development. Furthermore, the socio–cultural background of Bhutan, with its Buddhist value systems, is largely responsible for the country's overall well-being. However, he also points out the disparities between men and women, manifest in the literacy rates and in the gender division of labor, which is further exposed in the water sector.

Notions of Well-being

Well-being is a relational and dynamic concept and is widely debated. If we trace the term, we find its usage as "welfare" (in the 14th century) meaning happiness and prosperity. The shift was slowly seen in the 20th century when it began to be understood as assessment and provision of needs, as in a welfare state, acquiring an objective interpretation or a measurable definition. Later on, however, with discourses around participation, freedom, agency and new ways of understanding poverty gained currency, and charted the course for an older interpretation of well-being as happiness which was not necessarily measurable (Gough et al. 2007). Well-being, therefore, seems to be more of an overarching concept, accommodating different meanings and interpretations and is both an outcome and a process, determined by a constant interplay of social, political, economic and cultural processes. Conceptions of well-being are shaped by its dominant notions (mainly economic), focusing on the material resources that people own, utilize and dispose, measured in terms of Gross Domestic Product (GDP) or national per capita incomes that are often highly skewed averages.

From Income Poverty to Human Development

Development has historically been equated with material well-being and has been measured in terms of increases in GDP and Gross National Product (GNP). In recent times, however development is being recognized as an organized pursuit of well-being (Sen 1993; Chambers 1997; Gasper 2004; Gough et al. 2007). This calls for broadening the definition from its economic perception to an understanding of human well-being to encompass ideas of participation and freedom. Poverty or an income analysis is therefore not sufficient to assess the agency of people and the meanings or values they attach to their social and material resources. The transition from an income poverty discourse to a human development and well-being discourse has been a long and tortuous one. Amartya Sen's work has been considered as a prominent body of work. He disputed that command over property would suffice in assessing poverty and the well-being of people. Resources and agency should be expressed in terms of not just economic vulnerability but also political and social vulnerability. This prompts a richer analysis of the way people use their resources to mitigate poverty (Sen 1981). Later scholars, such as Martha Nussbaum, have challenged the income–poverty discourse by drawing attention to emotions, expression, imagination and creativity, among others. Along with Amartya Sen, in the book *Quality of Life* (1993), she argued against the traditional utilitarian views that see development, purely, in terms of economic growth and poverty as income deprivation. She expressed that poverty is often a result of capability deprivation. For example, to be able to participate in politics or have caring relationships, among others, are useful indicators to assess poverty and not incomes and economic growth alone.

Basic Needs to Well-being

If we trace the history of development in the South Asian region, we see that the transition from a basic needs approach to the human well-being approach has come a full circle, but through different theoretical and political arguments. In the pre-1990s, the basic needs approach was dominant and included fulfilling the needs of food, clothing and shelter for all. The approach gained momentum in the mid-1970s but was not popular among the diverse strands of thought. The rise of the neoliberal agenda dominated development thinking in the late 1980s, thus bringing the discourse of needs vs wants that can be met through emerging and growing markets. Around the same time, the post-modern discourse too was playing a significant role in shaping the political demands of the day–they argued that the basic needs approach was based on an arbitrary and monolithic understanding of peoples' needs. Eventually, the basic needs agenda was abandoned.

However, after 1995, basic needs came back in their new avatar of Millennium Development Goals (MDGs)–adopted in the General Assembly in 2000–bringing attention to development indicators such as infant mortality, access to water and education, among others. This approach was revived to address the rising instances of poverty and malnourishment, increasing number of people with declining access to basic amenities like water and food, increasing violence against women, resulting in increased instances of violence against women, Dalits (Scheduled Castes) and *adivasis*

(Scheduled Tribes). The different discourses that challenge these notions of development, therefore, become important. For example, development ethics (Gasper 2004) or feminist discourses–which talk of a need for altered value systems and new ethics as the very nature of development (who is benefited and in what way)–are important in assessing the outcomes of development. They take us beyond an analysis of the material or the tangible aspects of development to the more intangible ones that hold different values and meanings to people–therefore, going beyond increased incomes and improved calorie intake to for example happiness, freedom.

Gender, Resources, Agency, and Well-being

Assessing resource empowerment and well-being achievement is a very complex task, since resources accessed (land, water and credit) or controlled may not necessarily translate into agency and achievement for women or other disadvantaged groups. Various studies (Agrawal 1994; Mukhopadhyay 1998) have shown that translating access to a meaningful outcome is determined by various social, cultural and other determinants. Resource access or control has, therefore, been equated with a say in decision-making. An important point gleaned from the literature on land rights is the need to assess the agency and choice that women can exercise–as a result of control over access to a material resource or to realize the potential of that resource (Kabeer 1999).

As we have seen, discussions on well-being are located at different points on a spectrum ranging from a very liberal notion to one which engages with the social, cultural and political processes that affect well-being. This brings the concept of agency to the fore since it is understood to play a critical role in achieving well-being. However, it is also important to remember that the process of achieving well-being is also closely related to structural inequities that exist in society. Frameworks that overemphasize the agency of the poor often gloss over the role of the structure in constraining agency and how structure contributes to the poverty of their agency (Gough et al. 2007).

The concept of well-being is useful when it comes to assessing the impact on the disadvantaged groups. Understood in its broader sense of access and control over tangible and non-tangible resources and the agency to actualize these resources for meaningful outcomes thus becomes a useful starting point. In this chapter, I have used this broader meaning to discuss and contextualize it in a few micro-settings in the water sector.

Well-being in Water

In this section, it is important to consider how the previously-described literature and understanding translates into the water sector, particularly for women belonging to different castes, tribes or race. One of the important components of IWRM is to integrate the interests of women in the context of water. Since the 1990s, therefore, gender mainstreaming gained currency in water sector planning as international organizations recognized the need for mainstreaming in this sector (GWA and UNDP 2006). Globally, various policy initiatives, training programs and activities were launched to bring visibility to the gender question in an otherwise technocentric resource like water. But the initial enthusiasm around this discourse seems to have waned and the time has now come to ask some hard questions and take stock of issues in gender and water in the last two decades.

Competing discourses of human right to water, equity in distribution, sustainable use that recognize the value of the social, cultural and political dimensions along with technical, economic and institutional ones provide a useful starting point to understand well-being in water. In terms of practice, however, these discourses largely understand well-being as improved health, hygiene and overall prosperity at the household level. From the gender perspective, these impacts have been seen as the amount of time saved and hardship reduced in the collection of domestic water. As far as productive water is concerned, improved irrigation facility has led to improved agricultural productivity–thereby benefiting all members of the household and community equally.

However, the preceding analysis misses out on women's agency to translate access to resources into meaningful tangible outcomes for themselves. Moreover, it also misses the meanings that people attach to water as a resource and its bargaining capacity to negotiate for better social, political and economic outcomes–what can be referred to as the intangible or non-measurable outcomes. The human well-being approach, if discussed beyond incomes, health and hygiene benefits, will further help us understand the diverse meanings that people attach to water. The questions that we need to start asking are: whether there is access to or control over water, whether choices and conditions are available and to whom to realize this access?

Gender, Water and Well-being

There are systematic disparities in the freedom that men and women enjoy, across different social groups, which are not reducible to incomes or resources although there is a significant difference in the ownership and access to these as well. Added to this are the differential benefits in terms of care, attention, division of labor, education and other opportunities. These inequalities cannot be captured by the division of income to each member but can be understood through the use of the resource. It needs to be seen whether the use of resource has changed the ability of the person to effectively function as a social being. For example, with regard to women, it would mean altered relations with men and the community or the household.

In the context of water, discrimination against women (from diverse social groups) is manifested through access to water and their representation in water-related institutions. It comes in the form of excluding women and Dalits or ethnic groups from accessing water based on land ownership. Also, women's unpaid work around water is hardly accounted for, whether in irrigation or drinking water. Drawing on some of the inferences of fieldwork conducted in in Maharashtra, Gujarat[1] and Andhra Pradesh, this chapter looks at the impact of water programs and policies on the well-being of women. In Maharashtra and Gujarat, women's participation was studied in the context of the sector reforms process in the water sector, which promoted the formation of decentralized water institutions and people's participation in decision-making. In Maharashtra, we looked at Jalswarajya (funded by the World Bank), styled on Swajaldhara–a flagship program of the Government of India where community contribution (in capital cost) is 10 percent and the state contribution is 90 percent. At Jalswarajya, we mainly studied women's participation in the Village Water and Sanitation Committees (VWSCs), popularly known as *pani samitis*, as they are expected to have 50-percent representation of women. In the irrigation sector, we looked at the Water Users Associations (WUAs), which are water-related institutions in surface irrigation and have about 33-percent representation of women, owing to the newly introduced legislation.[2] In Gujarat, the study focused on the drinking water projects of (*a*) the Ghogha project, funded by Dutch bi-lateral aid;[3] (*b*) Earthquake Reconstruction and Rehabilitation (ERR) project in Kutch;[4] and (*c*) NGO-facilitated and community-led water interventions within the framework of sector reforms[5] in these decentralized institutions. In Andhra Pradesh, the study was done in Warangal district where water interventions had been carried out by Modern Architects of Rural India (MARI), an NGO working in the district on a range of development issues. The main objective of the study was to look at how and whether water interventions are meant to reduce the impact of poverty on women.

In the context of well-being, the broader areas of study included (*a*) equity in access and voice; (*b*) beneficial impacts on health and hygiene; (*c*) beneficial impacts on "drudgery reduction" and "time saving;" (*d*) freedom of choice in the use of water and associated technology; and (*e*) freedom in articulating gender and equity concerns and a social justice agenda.

Equity in Access and Representation

Access to drinking water has to be universally available, irrespective of land holding, caste or economic status. Water for women, therefore, becomes part of this understanding. However, both in Gujarat and Maharashtra, we see that access to water is inequitable in terms of caste and economic status. The new water paradigm is increasingly talking a language of pricing and the ability to pay– that is, access to assured and safe water is dependent on one's capacity to pay for it. Consequently, the worst affected groups are the economically disadvantaged, the Dalits and the *adivasis*, while affecting women twofold. Increasingly, new schemes also encourage private connections over public stand-posts to make cost recovery easier, thereby making it less and less affordable for the poor. Therefore, access is differential–depending on the physical, political and socio-economic location of the communities. These schemes ostensibly promote transparency and equal representation in voice (among others) and, accordingly, have managed to pull together some kind of token representation across caste, tribes and underprivileged groups. However, these representatives are hardly able to articulate the concerns of those they represent, either because they are co-opted by the dominant groups or simply because their voice is never heard.

Equity is not a fixed concept and operates at many levels. In the context of public sector irrigation, equity within the command area of the irrigation project refers to every piece of land that has access to water. Apart from this, everyone in the command should have access to minimum water, irrespective of his/her land holding and women, in particular, should have access to water.

Politics of Exclusion

Our data points to different ways in which inequities across social groups are manifested in the irrigation sector. At the very first level, we see a politics of exclusion based on land ownership in the command areas of the canal systems. A large number of women belonging to landed families within command areas, men and women belonging to landless families and landed families outside the command area are automatically excluded from the process.

A caste-based analysis in the same study shows that Dalits or the Scheduled Castes (SC) are rarely land owners in the command areas of irrigation systems and that the proportion of Dalit farmers in the WUAs is not commensurate with the Dalit population of that village. Managing Committees (MC) are the decision-making bodies of the WUAs. The minimum number of MC members is usually nine but can extend to 11–13 depending on the size of the WUA membership. However, representation to the MC is highly influenced by caste. Data on women's membership to decision-making committees shows that only one society had a woman representative from the Other Backward Castes (OBC). The present Act has made it mandatory to have at least three women on the MC from each of the reaches of the canal. In most cases, these women belonged to the upper castes and were instituted as proxies to the influential and powerful men in the WUAs.

Beneficial Impacts on Health and Hygiene and on "Drudgery Reduction" and "Time Saving"

Beneficial impacts of improved water availability have been widely discussed, primarily in the context of improved health and hygiene of the household and in terms of "time saved" and "drudgery reduced" for women and/or those who collect water for domestic use. In the irrigation sector, improved access to water is measured in terms of improved productivity, better cash incomes and diverse cropping patterns (equally distributed among members), all of which are assumed to translate into better nutrition and general well-being in the household.

In villages in Maharashtra, where the schemes were completed, more than 45 percent women cited "time saving" as the main impact followed by a positive impact on the health of the family, which also contributed to the overall cleanliness of the village and home. Women said that this saved time could be used for other activities, such as doing additional household and agricultural work. This shows that most women do not perceive time saved as time for leisure. In the Andhra Pradesh study, women did say that water availability had changed their lives. One of them (in pers. comm.) said, "Yes, with increased water we have taken up vegetables apart from cotton and paddy, which cut down the money spent on buying vegetables. The family eats more fresh vegetables and this has a positive impact on the health of the family including that of women."

Increased incomes through sale of crops have been variously used by different households which have a lot of intra-household variations. These include a range of activities, such as purchase of gadgets like televisions, cell phones, two wheelers, house repairs or health needs. Many have also used this cash to repay old loans or buy crop requirements, such as seed and fertilizers. A few of them have also invested in toilet construction or towards buying food grain for the family. The income from crops is spent on purchasing motorcycle for men in the family. Some amount is spent in house repairs, laying of asbestos sheet roofing for the house, or constructing a toilet.

Increased cropping has also meant better food availability for the house. Almost 50 percent of them saw a marked difference in their food availability and also their own food intake. The diversity of food available also increased. Vegetables and fruits along with pulses and cereals were now the diet pattern of these households. The data shows that while crop diversity and productivity has increased, the cash does not necessarily translate into positive outcomes for women. Most of the cash is spent on buying consumer items used largely by the men of the household. Although the overall food availability has increased, it is difficult to assess how much of this is accessed by women. Therefore, an evaluation of the intra-household distribution of increased food is often complex.

Increased water availability for agriculture has meant that both men and women have to engage in additional work on the field. Traditional tasks of weeding, harvesting and engaging labor continue to be done by women, thereby increasing their drudgery. In Andhra Pradesh, irrigation led to improved availability of fodder. Consequently, the time spent by children grazing cattle has reduced, improving their attendance in school. On the other hand, the women have said that they now faced the entire burden of managing the cattle.

Public Sphere Participation as Well-being

Recent water programs are bringing women in the public sphere through the different quotas introduced in the committees. Often, this results in additional work for women without any change in their conventional work patterns. The question of opportunity costs is often difficult to assess as women rarely find an opportunity to go beyond the realm of household work. Opportunity and work burdens hold different meanings for women coming from different locations. For most of them, however, this was the first opportunity to participate in public matters and, hence, they did not perceive this as a burden or an additional cost. Women had to put in additional time to be able to participate in these micro-level processes. Most of the women acknowledged another woman's support in this regard or said that they worked hard in both the house and in the public sphere. This meant that there were competing demands on their time.

Freedom of Choice in Use of Water and Associated Technology

Naila Kabeer (1999) has done an extensive mapping of how conditions and consequences of choice determine the outcomes of resource access. For instance, political, social and cultural contexts play an important role in women's ability to decide upon the use of resources. One of the main areas that we wanted to investigate was the impact of a water intervention, such as watershed development or an irrigation project or a drinking water facility, on women's choices and freedom. Our studies in Maharashtra and Andhra Pradesh showed that women's options for choice and autonomy do not improve with new water interventions, since most decisions related to crop preferences or uses of water are taken by the men of the household. Even a membership to sugar factories, the choice of water infrastructure or technology largely rests with the men.

In Andhra Pradesh, as far as decision-making is concerned, we see that the male choice still dominates and though women have started expressing their interests, they do not differ drastically from the mainstream choices–which are governed by soil conditions, market situations, availability of seed, water or other resources, and mainly by the prevailing social norms.

Freedom in Articulating Gender and Equity Concerns and a Social Justice Agenda

Women's representation in water committees often does not translate in better articulation of their own concerns. In the decentralization study, one of the areas that we discussed with women was whether they could articulate the concerns that they thought were important. Among the domestic water committee members, more than 50 percent of the women said that they could speak out their concerns in these meetings. This overall positive response is encouraging and speaks, to some extent, about the spaces created by the scheme, which go beyond quotas. Here, the role of the *mahila gram sabhas* (all women's village meetings) and separate meetings for women are seen as important, thereby indicating that social context and a political will could go a long way in creating the ground for well-being. However, on further probing about awareness and facility of raising issues around water, during the *pani samiti* meetings, we got a caste-differentiated understanding which told us that Dalit women were among those who were not able to discuss their problems.

However, the findings in the irrigation sector are quite different. Very few women are eligible to become members simply because they do not own land. Those who owned land, in the command areas of the irrigation projects, were seemingly unaware that they were members of the MC. They were also not aware of their role in the WUA and, therefore, had not attended any of the meetings.

Way forward

The different studies discussed in this chapter show that though water programs do bring in an overall improvement in household health and hygiene and enhance the economic status, this does not necessarily translate as well-being for the women of the household or the deprived sections within the community. Women continue performing the old tasks in addition to the new labor, the result of a new water program. The time saved from walking long distances to fetch water is usually invested in additional housework or growing more crops and/or running allied activities, such as livestock and dairy, apart from carrying out committee duties. When placed against the background of patriarchal forms of division of work and inequalities, the new water programs are hardly able to address the well-being concerns of women. Laws or policies in favor of the poor or women cannot circumvent social structures, norms and customs. Most of our experiences with progressive law have shown that social traditions and customs have largely prevailed and prevented democratic participation and freedom of expression of women.

The well-being framework discussed in this chapter helps us look at the different impacts of water resource programs that go beyond the impacts of health, hygiene and economic prosperity. It looks at the different values and ethics that are upheld in planning for interventions and outcomes. Unless these value systems and ethics are questioned and altered, well-being as defined here would not be attained. The way forward, thus, lies in re-imagining communities, the status of women and, independently, the water sector which has its own set of value systems and ethics.

Notes

1. A two-year study was conducted on understanding decentralized water management and its impact on women's empowerment and participation in water governance across nine districts in Maharashtra and six in Gujarat, from 2006 to 2008. The study titled "Water Rights as Women's Rights? Assessing the Scope for Women's Empowerment through Decentralized Water Governance in Maharashtra and Gujarat," was jointly carried out by SOPPECOM, Pune, Utthan, Ahmedabad, and Tata Institute of Social Sciences, Mumbai, and supported by International Development Research Centre (IDRC) as part of its global research on decentralization and women. In Andhra Pradesh, the study was carried out in one district and looked at the interface between gender, water and poverty. It was a collaborative effort between Gujarat Institute of Development Research (GIDR),

Ahmedabad, and SOPPECOM, Pune, and was supported by WaterAid India. Titled "Water Poverty and Gender: Understanding the Interface and Drawing Policy Implications," the study was conducted in 2010.

2. In 2005, the Maharashtra Management of Irrigation Systems by Farmers Act was introduced. Among the several clauses to improve participation of farmers it also included women's participation in decision-making committees.

3. The overall purpose of the project is to develop, integrating with water resources management, improved, safe, reliable and sustainable drinking water and environmental sanitation provisions in 81 villages and one town of Bhavnagar district in Gujarat, the facilities of which will be community-owned and-managed through the local *pani samitis*.

4. The ERR project was launched in 2003 after the 2001 earthquake that destroyed the Kutch district of Gujarat. It was aimed to restore and develop the water facilities in 1255 earthquake-affected villages in Kutch. These projects are part of a demand-responsive strategy where the role of the community was seen as critical in water management.

5. Swajaldhara and similar such schemes had a strong NGO component in them. Government called on NGOs as partners in the implementation of water programs.

References

Agarwal, B. 1994. *A Field of One's Own: Gender and Land Rights in South Asia.* Cambridge, UK; New York, USA: Cambridge University Press.

Barlow, M. and T. Clarke. 2002. *Blue Gold: The Fight to Stop the Corporate Theft of the World's Water.* New York: New Press.

Chambers, R. 1997. "Responsible Well-being: A Personal Agenda for Development," *World Development,* 25(11): 1743–54.

Gasper, D. 2004. *The Ethics of Development: From Economism to Human Development.* Edinburgh: Edinburgh University Press.

Gujarat Institute of Development Research (GIDR) and Society for Promoting Participative Ecosystem Management (SOPPECOM). 2010. *Water, Poverty and Gender: Understanding the Interface and Drawing Policy Implications.* Final report of a one-year study in Andhra Pradesh and Madhya Pradesh supported by WaterAid India.

Gough, Ian, Ian McGregor and Laura Camfield. 2007. "Theorizing Well-being in International Development," in Ian Gough and J. Allister McGregor (eds), Well-being in Developing Countries: From Theory to Research, pp. 3–44. Cambridge; New York: Cambridge University Press.

Gender and Water Alliance (GWA) and UNDP. 2006. *Resource Guide: Mainstreaming Gender in Water Management,* http://www.washinsschools.info/docseanch/title/124357 (accessed April 23, 2013).

Global Water Partnership (GWP). 2010. "What is IWRM?", ww.gwp.org/The–Challenge/What-is-IWRM/(accessed April 23, 2013).

Harris, L. 2009. "Gender and Emergent Water Governance," *Gender, Place & Culture*, 16(4): 387–408.

Kabeer, Naila. 1999. "The Conditions and Consequences of Choice: Reflections on the Measurement of Women's Empowerment." Discussion Paper No. 108, United Nations Research Institute for Social Development (UNRISD).

Kulkarni S., S. Ahmed, C. Datar, S. Bhat, and Y. Mathur. 2009. *Water Rights as Women's Rights: Empowering Women through Decentralized Water Governance in Maharashtra and Gujarat.* Final report of study supported by IDRC, SA.

Mollinga, Peter, Ajaya Dixit and Kusum Athukorala. 2006. *Integrated Water Resources Management: Global Theory, Emerging Practice, and Local Needs.* Water in South Asia, Vol. 1. New Delhi; Thousand Oaks: Sage Publications.

Mukhopadhayay, M. 1998. *Legally Dispossessed: Gender, Identity and the Process of Law.* Calcutta: Stree Publications.

Nussbaum, M. 2000. *Women and Human Development: The Capabilities Approach.* Cambridge: Cambridge University Press.

Seager, Joni. 2010. "Gender and Water: A Good Rhetoric but It Doesn't Count," *Geoforum*, 41(2010): 1–3, Geoforum Journal Homepage. Also available online at http://www.journals.elsevier.com/geoforum (accessed March 25, 2013).

Sen, Amartya. 1981. *Poverty and Famines: An Essay on Entitlement and Deprivation.* Oxford: Clarendon Press.

———. 1992. *Inequality Reexamined.* New Delhi: Oxford University Press.

———. 1993. "Capability and Well-being," in Amartya Sen and Martha Nussbaum (eds), *The Quality of Life.* Oxford: Clarendon Press.

Swyngedouw, E. 2006. "Power, Water and Money: Exploring the Nexus." Human Development Report Office Occasional Paper, United Nations Development Programme (UNDP).

United Nations Development Programme. 2006. Human Development Report: Water for Life. New York: UNDP.

2

Women, Water, Caste, and Gender

The Rhetoric of Reform in India's Drinking Water Sector*

Deepa Joshi

⬤

"We are *domnis* today!" say caste Hindu menstruating women in rural Kumaon villages in India, explaining their ritual monthly seclusion from the water springs, until they are purified on the 5th or the 7th day (in pers. comm., Joshi 2002)

A universal patriarchy creates a structural and symbolic inequality between women and men (Molyneux 2001); yet race, caste, colour, religion and class deny a universality of gender inequality (Mohanty 1991). The issue of who carries water home and how is central to gender–water relations, yet this basic fact is often overlooked both in policies and discourses around water and equity (O'Reilly et al. 2009). On the contrary, projects make "normative assumptions about women's [water] roles, actively reinforce and deepen gender divides . . . through an increasing feminization of [water] responsibilities and obligations" (Molyneux 2007: 231). As Sierra Tamang writes, the politics of developing poor, Third World women conveniently "effaces ethnic, religious gendered differences among heterogeneous communities and constructs an [illusory] category of women–all generically poor, backward and needy yet willing, capable and committed to making projects work" (2002: 317). That such generalizations occur also in India is of immense concern given the existing reality that already disparities by caste and do not allow for a simplistic categorization of gender. According to Dip Kapoor, ". . . any attempt to address gender inequity in India [seriously] needs to consider the socio-cultural vector of a gendered, caste-based discrimination against untouchable Dalit women" (2007: 609).

Several factors have contributed to a misinterpretation and, subsequent, depoliticization of gender in development. In this chapter, I discuss the early eco-feminist emphasis on women's innate relationship with the environment (Mies and Shiva 1993) and its translation in development, to the positioning of women as an attractive source of willing labour for repairing the environment (Kabeer 1994). In India, the misinterpretation of gender as women was furthered by a contextual segregation of disparities by gender and caste. Subsequent policies have failed to take into account that caste, class and gender often intersect to position Dalit women at the bottom of the social hierarchy (Govinda 2007).

Water equity discourses often overlook complex gender–caste–domestic water issues and make vague, often unsubstantiated, assumptions when eulogizing the equity outcomes of one institutional approach over another. In this context, the recent handing over of drinking water responsibilities, from state water institutions, to diverse combinations of State, private sector and community institutions in India, has been argued by many as a historic loss of equity. Phillipe Cullet (2009) and others make insightful analyses of the negative outcomes of these changes in policy for those marginalized by class and caste. Yet, there is often a debilitating silence in these analyses about women, especially Dalit women's challenges in carrying water home. However, contrary to the previously mentioned concerns relating to the loss of equity, David Mosse advocates for the "twenty-first-century neoliberal reverse 'rolling-back' of the state claim that the process allows for a 'revival' of community [read women] water management" (2008: 946). Indeed, the recent neoliberal strategies are promoted to be strategically promising towards equity, especially 'liberalizing' to women, where pricing mechanisms . . . promise a free palate of choices to satisfy [everyone's] all needs . . . ' (Ahlers and Zwarteveen 2009: 413). Far from compromising on equity, recent reforms in India are premised as enabling "women to [pay for and obtain water they need or demand and] throw away once and for all the pots that they use to carry water" (Sangameswaran 2010: 66).

This chapter compares discourses around policy reforms in the drinking water sector (in India) to ground realities of diverse women–water experiences to argue on the superficiality of gender claims and blames in evolving policies, and illustrates that the equity benefits of welfare-based and supply-led approaches as well as the neoliberal one have shown a poor understanding of complex gender–caste–water realities.

Caste, Gender and an Unequal Water Order in India

Deepa Joshi and Ben Fawcett's (2006) analysis of the complex intersection of caste, gender and water in India identified two issues. First, there are distinct differences in the experiences of caste- and gender-based inequity, even though both are outcomes of the same principles. Second, disparities of caste and gender converge in complex intersections, making it impossible to neatly segregate them. For example, Kapoor (2007) reports how Dalit women, burdened both by their sexuality and caste, interpret sexual atrocities by caste Hindu men as acts of caste dominance. Yet, few agencies in India, formal or informal, recognize the complexities of caste and gender intersections (Govinda 2009). I outline below women's experiences around domestic water in the Kumaon region illustrating how these experiences are nuanced locally for Dalit women.

Women, Water and Gender

Focusing on local perceptions of women's Nature–culture interactions (in the Kumaon region where my research is based), Shubhra Gururani (2010) argues that neither eco-feminist "romanticization of women's association with Nature" nor the work of feminist ecologists–who expounded that women–Nature responsibilities are imposed by patriarchal local cultures–explain the reality in the local settings of the Kumaon villages. Gururani observed that activities relating women and Nature, such as carrying fodder or water home, are indeed constraining but they also provide women with unique outlets to establish their worth or value within the home as well as build and nurture social relations outside the restrictive boundaries of the home.

In my interactions in the region, however, I observed waves of coercion, anger and finally, a resignation amongst women in the gradual but distinct gendering of domestic water responsibilities. As young children, both boys and girls fetch water but a sharp gendering of water responsibilities takes place during adolescence. Young, unmarried girls were the angriest and most critical of how they were drawn to the numerous tasks at home, while their brothers and other males of the same age were released from the same because these duties were considered to be domestic and feminine. Only unique personal situations result in young men performing domestic water-related work. And when adult men assist their wives at home, the local constructs of male hegemony become even more severely violated and local culture does not readily support this.

However, the burden of constraining adolescence and overwhelming domestic responsibilities faced by young girls becomes a pleasant memory for the elder, married women. The cementing of water responsibilities in these mountain villages happens on marriage. A ritualistic visit to the family spring, the ceremonial fetching of water home (the day after the wedding) makes this task, from here on, *binding*. These tasks continue for women even after their bodies are bent with age and physical exertion. In my discussions locally, the only women who cherished every opportunity of getting out of the home, including fetching the heaviest pots of water, were young married women, especially those whose husbands were migrants and whose social status as young brides equated to being glorified slaves in their husbands' homes. Nonetheless, these same women talked fervently of visiting their maternal homes, of being able to forego (for a few days at least), the burdens of their now binding domestic responsibilities.

The point I want to emphasize here is that domestic water responsibilities are indeed imposed and if women see these tasks as uniquely distinctive [or rewarding?], it is because they know no other option. However, what I also observed was that fetching water is not equally perilous to all women. Age, social status and, above all, caste denies a commonality of "women's water burdens."

Caste–Gender Complexities and Water

At the start of my research in 1998, I had planned to visit water-abundant and water-scarce villages in the Kumaon region in order to assess, what I assumed would be differing aspects of gender–water–caste dimensions. My naiveté of caste–gender complexities was revealed in the very first village I visited. The residents, especially the upper-caste Khanka Kshatriya women, of Chuni village (in the mountains of Uttarakhand) reported water-abundance and claimed that it was due to the Water Goddess (*Jal Devi*), revered in a temple located at the foot of the village. She is praised for ensuring that water flows around the year in the springs and along the narrow irrigation canals (*guls*)–"Whatever else our problems we are water-*lords* [sic] here." In the same village, however, the Dalit Agari women said (in pers. comm., Joshi 2002):

> Ask us what water scarcity is? It is not being able to bathe in the summer heat, after toiling in the fields. It is to reuse water, used in washing vegetables and rice to wash utensils, to use this soapy and dirty water again to wash clothes and then feed it to the buffaloes. Water scarcity is to sit up the whole night filling a container, glass by glass, as it trickles into our one small spring. We often don't wash the utensils and just wipe them with a cloth. We feel so dirty and unclean in the summer. We do not wash our clothes for weeks, just rinsing them with a little water. These people say we are dirty and we stink. But how can we be clean without water?

Access to water sources by the Dalits is linked to fears around pollution. The small, flowing agricultural *guls* and freshwater underground springs (*naulas*) are not accessible to these Agari Dalits. Historically, Dalits live in hamlets distanced from the main village and the primary water sources, especially the temple areas where the main springs are situated. "Good Dalits" are those who keep away from such sacred spaces and water sources used by the upper castes. Located away from this sanctified water-abundant main village, eight Dalit households share one small spring while, nearby, a single spring is used by one Khanka family. The sanctity of the Khanka *naula* would be defiled and its water polluted by the touch of the Agaris. Therefore, they must wait for water that the neighboring Khanka family provides at will while demanding and securing various obligations (Joshi 2002).

The water sources are also forbidden to the upper-caste Hindu women of Chuni in their cyclically "impure condition." When the water level suddenly drops (particularly in very dry months), popular rumours of the water being polluted abound and various purification ceremonies are performed to cast aside the pollution. Conversely, these women are always blamed for having entered these sacred spaces while menstruating or when unclean. To blame the Dalits, in this situation, would justify their access, even in hiding. "We are *domnis*¹ today!"–is how upper-caste Hindu, menstruating women often announce their ritual, monthly seclusion from the temple or the water springs, until they are purified on the 5ᵗʰ or the 7ᵗʰ day (after the first day of the bleeding). Menstruating Dalit women are, therefore, doubly impure!

In my discussions with the Dalit households in Chuni, both men and women spoke of and illustrated the blatant acts of caste dominance. Men spoke of their continued "lesser men" status–how even [upper-caste] women younger to them in age look down upon them, offer them food or water from a distance, require them to wash their own cups or plates, among others. Dalit women spoke most of the practical challenges of finding and carrying water home in water-abundant Chuni–of the perils of *stealing* the cool, sweet water in the *naulas* on hot summer days; how the consequences of stealing are minimized by sending out young children to "steal" water and then beating them up severely in a faked guilt (if caught); how religious ceremonies are performed by upper-caste families to "purify" the *naula* (if discovered); and of the long and "hard" costs imposed if caught "stealing and polluting" water. Nandi Devi Agari asked, "What can we do? Will you come with me to the Khankas' *naula* and I will take out water in the open (in daylight, in everybody's knowledge)?" However, she realized the futility of her idea. "You sympathize with me, but you are here today and gone tomorrow" (in pers. comm., Joshi 2002). Conflict and defiance are not her preferred solutions. My following discussion outlines how water governance and management has changed several hands in Chuni since India's independence, but there has been little progress in Dalit women's access to water.

Gender Asymmetries in India's Drinking Water Sector

It is ironical that in India–where gender, caste and water combine to structure an unequal social order–early theoretical discourses around gender and the environment romanticized the collective of communities as well as women's unique relations with Nature. Shiva's (1988) evocation of a divine femininity and the inherent links between women, Nature and nurturing, especially in the Kumaon Himalayas, was complemented well by traditional environmentalism praise for a mutually beneficial entwining of Hindu ritualism and conservative Hindu culture (Agarwal and Narain 1997; Rawat and Sah 2009). Praising the technical ingenuity and common ownership of the *naulas*, Anil Agarwal, the activist-environmentalist, credited with have redefined environmental problems through the eyes of the poor people (especially women) claimed that "there were few caste and class barriers in the Himalayan villages which prevented people from working together as a community" (1985:13). Analytical critiques of "traditional environmentalism" identify that these systems were often far from equitable (Sinha et al. 1997; Mosse 2008). Mosse also argues that there are few "real" accounts of these traditional histories and the romanticizing is often based on "veins of myth and memory" (2008: 241).

Joshi and Fawcett (2006) describe in extensive detail people's lived experiences in the mountain villages in Kumaon, which rarely match the romanticized narratives of village collectives. Women in Chuni could not recall a single period in history when women were consulted or involved in the planning and design of water systems. Similarly, Dalit households cannot recall when they were ever considered an inclusive part of the village community. Indeed, water was and still is the medium of exclusion. The villagers, both Dalits and non-Dalits, corroborated the fact that after Dalit artisans built the traditional *naulas*, these structures were sanctified and purified [sic] of their "polluting touch." Unequal access to water in the village continues, legitimized by religion and caste, and the Dalits (in their minority) are unable to challenge this deep-rooted social inequality.

The inconsistencies in these blames and claims around collective action, women and the environment are important because they have allowed, at least in India's drinking water sector, a continued reproduction of gender myths in (later official and now contemporary) water policy (Sinha et al. 1997).

State Ownership and the Management of Water

The effectiveness and appropriateness of the state's role in water management is much disputed. Scholars like Cullet (2009) conjure notions of a welfare state and its official intent to address the fundamental right to water. Others like Shiva (2001) find little to applaud in the state, ". . . institutions constituted to achieve colonial economic and political ambition, and now intervening through policies, rules, laws, investment, and technology to facilitate privatization and globalization." In this chapter, I have discussed my observations of the late 1990s–a limited understanding and application of complex gender and caste concerns in official implementing and policy organizations.

In India's official policy, caste was prioritized early and, consequently, conceptualized as separate and distinct from gender. A clause that prioritized coverage of drinking water supply to Dalits and Sudras, the next higher order caste group in the caste ladder (both identified officially as Scheduled Castes), was applied in policy in the early 1980s. It was meant to address what was identified as a debilitating lack of access to drinking water for the lower castes, said to be furthered by the deep-rooted caste connivance between local officials and villagers of identical castes (Agarwal 1981). However, no strategies were identified to challenge this issue. According to Tiwari (2006),[2] quoting official data from the National 2002 Census, it is not a surprise that, after two decades of prioritized access, Scheduled Castes (especially Dalit women) continue to have reduced access to and also travel significantly longer to fetch drinking water. In the course of my research, a walk through the different hamlets in Chuni revealed numerous accounts of continued connivance between upper-caste villagers and officials in the misuse of official funds. In an ironic twist to policy, official funds had been consistently used to divert the meagre water resources of Dalit households to meet the emerging needs of the dominant upper-caste community, mainly inconveniencing Dalit women, placed at the lowest end of the class, caste and gender hierarchies.

By the late 1980s, "gender" was formally incorporated in the drinking water policy in India. However, policy-makers in New Delhi understood the purpose of including an undifferentiated group of "women" as the means to address efficiency–"[W]omen being the biggest users, collectors and handlers of water have a major role to play in rural water management as caretakers, health educators and hand-pump mechanics" (Ghosh 1989). However, since progress in meeting targets was defined by (*a*) funds spent in delivering water, and (*b*) hand pumps installed and working–with data disaggregated by general and Scheduled Caste communities, there was little need (fortunately) for field-level engineers to involve women in managing water, according to the limiting ways defined in the policy.

In the rolling over of the state to neoliberal arrangements in delivering water, Radhika Govinda (2009) writes of the politics of prioritization of gender over caste, which complemented the neoliberal focus on the revival of women's roles in water management. In the following section, I discuss how the isolation of caste and gender considerations has served to deny Dalit women the privileges accorded by the affirmations of both gender and caste.

From Supply to Demand-led Interventions: Neoliberal Approaches to Domestic Water Supply

In the 1990s, internal and external reviews were beginning to highlight the poor performance of state water supply institutions. After being disenchanted with the supply-driven, public-utility-managed policies, rural water supply was influenced significantly by a review of the sector performed by the World Bank in 1998– which gave a broad-brush analysis of the faults in the supply-driven approach. However, rather than making an attempt to analyse or correct these drawbacks, the report called for "[a]n urgent shift from supply-driven to demand-oriented approaches . . . an explicit engagement of non-government stakeholders in sector activities– with an aim to achieving financial viability of service delivery" (World Bank 1998). Gender advocates operating within the neoliberal framework propose multiple social and financial merits of a Demand Responsive Approach (DRA)–"The DRA takes into account that rich men, rich women, poor men and poor women may want different kinds of service. DRA allows user choices to guide key investment designs, thereby ensuring that services conform to what people want and are willing to pay for" (Dayal et al. 2000: 2).

Contrary to Cullet's (2009) concerns on the loss of a historic equity, Ahlers and Zwarteveen (2009) and Harris (2009) point out that there are many counterclaims on the immense possibilities for equity as an outcome of neoliberal approaches to water management. Consequently, I discuss how this re-positioning of women–as a community, in the reform agenda from user citizens to consumer clients–provides a much higher incentive to involving caste women (in making projects work) as well as ignores fractures between women caused by caste disparities. However, Cullet (2009) and several other scholars have elaborately illustrated the negative outcomes of the reform agenda for Dalits and the other poor marginalized families. Yet, there are few accounts of the impacts of these new policies on Dalit women. In the following section, I have discussed how the much eulogized Swajal Project, precursor to the recent neoliberal policy reforms in India's drinking water sector, reiterated entrenched gender–caste–water inequities, despite claims of gender and caste empowerment.

The Swajal Project

The Swajal Project (1996–2002) was planned and supported by the World Bank and managed by the Department of Rural Development, Government of Uttar Pradesh (Uttarakhand was later carved out from the state of Uttar Pradesh). Contrary to popular critiques, this project was, perhaps, the most articulate in emphasizing (at least, on paper), the terms of the neoliberal agenda: (*a*) equity, (*b*) empowerment, (*c*) caste and, (*d*) gender. Even as the project was barely underway, claims were being made about its innovative nature and "demonstrated" outcomes of financial and social sustainability–particularly, that in place of the overstaffed and corrupt, politicized state government implementing organizations, who practiced an inflexible technology, NGOs and private consultants would empower village communities, especially women (World Bank 1996).

The project's demand-responsive approach was based on underlying assumptions of community homogeneity and shared water needs and demands. The VWSCs would select scheme types, collect user fees, purchase materials, and arrange labor for construction. The goal of "women's empowerment" translated into 30 percent participation of (an undifferentiated mass of) women in the VWSCs (World Bank 1996). Educated women, especially, were prioritized as nominees for VWSC treasurers. Additionally, it was proposed that informal education and income-generation training would enable women to invest in and profit from the time saved in collecting water. Recognizing caste inequality, the project also specified 20 percent representation of Dalits in the VWSCs, but it was not specified whether this meant Dalit men or women.

There are several stories reporting the project's success in empowering women–told by partner institutions and their field staff; mentioned in World Bank project fact notes; anecdotes by official government visitors "impressed" by what they saw on field visits; and stories by consultant gender advocates–all emphasizing how the gender-sensitive structure of Swajal had benefited women the most. Some independent studies, such as the one by Prokopy (2004), also report high gains in terms of equal and appropriate access to water in the project areas. However, the data presented by Prokopy is not disaggregated by gender or caste, and does not explore pre- and post-project conditions of water scarcity and access.

In my observations, the stories unfolding in these mountain villages were significantly different from these claims of empowerment and equity. In Mala (the first village to implement the scheme), the communal standposts provided through the project proved to be convenient only for those households which could afford to pay an additional connection fee and divert water to private (individual) taps at home, allowing a 24-hour water supply. This had a significant impact on improving privacy and convenience for those women whose families could afford and chose to install a private connection. However, most of the poorer women still walked to collect water and the open design of the tap-stands meant that they still needed to rely on other alternative sites (better sheltered, isolated) for bathing and washing.

More importantly, while women in villages like Mala had been instrumental in making the project work, the dynamics of such engagement were rarely assessed in project documents. For instance, in Himtal village, members of the highly praised all-women committee said:

We participated because the men refused to do this voluntary work for no gains. It has been an enormous struggle to complete the project and to meet our various work demands at home and in the field. We are unable to change anything in the project design. We do not get any monetary compensation, not even to meet travel costs to the NGO and Project Management Offices. This is the price we women must pay for getting water (in pers. comm., Joshi 2002).

While project brochures showed VWSC women members proudly standing beside the new water delivery structures, women explained that their participation had not translated into taking decisions. For example, women (in Himtal) had never been asked to even consider questioning or changing the design of the open public stand posts–which was a predetermined project guideline. Furthermore, many upper-caste Hindu women did not come to these water points when menstruating, while Dalit women did not share communal water points with the upper caste households.

In the landmark project village Mala, Bina Devi, head of the lone Dalit single-adult household, was excluded from every single project benefit. Many reasons were provided for the exclusion–the upper caste VWSC members claimed that her home was outside the village boundary; others pointed out that as a head of a single-adult household she would be unable to attend meetings; some even "pointed" to her lack of interest in the project, given her inability to make the required payments. Senior staff at the NGO claimed to be unaware of this "anecdotal" fact. While the service agency staff rigorously monitored the technical infrastructure, there were no consultants to assess equitable participation. The field staff also admitted candidly, "To insist on her representation would have antagonized the dominant higher caste community in the village and hampered timely completion of the project" (in pers. comm., Joshi 2002). Ironically, the inaugurated scheme had taken over Bina Devi's water source and excluded her from accessing the piped and tapped water!

Entrenched caste divide frequently tested fragile gender gains. Local staff attributed the success of the timely project completion in Mala on its efficient treasurer, Maya Devi, an upper-caste woman whose family was among the wealthier ones in the village. Maya Devi is an active beneficiary and advocate of the local NGO; she also works as an informal teacher in the crèche run by the

NGO and is the first to be consulted for all NGO interventions. In interviews and conversations, she was also the most vocal in declaring that no wrong was done in excluding Bina Devi from project benefits. Such incidences were not anecdotal but indicative of systemic and complex caste-cohesive realities. The high profile of the Swajal Project in Himtal owed much to its president, Hema Devi (of the Kshatriya caste), who had relentlessly (and voluntarily) participated to achieve project demands. This included making a trip to distant Delhi for the purchase of cement and pipes, "I spent a terrifying night, when they stopped the truck on a lonely road and all the men got drunk" (in pers. comm., Joshi 2002). Though seen as an example of empowerment, Hema Devi found that, in fact, little had changed in the local social order. Midway through the project, her young son eloped with a married Brahmin woman of the same village. The dominant Brahmin community of Himtal did not tolerate this well; they blamed her "boldness" for this unfortunate situation, which they saw as a challenge to the existing social fabric of caste and gender hierarchy. Little blame was placed on the Brahmin girl's family involved but Hema Devi's family was ostracized for several months and she was asked to resign from managing the water project. When she tried to resist, the village leadership entangled her family in legal complications and she was subjected to enormous social and political harassment. No one came forward to assist her, not even the NGO staff who had promoted her "empowerment." Some of the women committee colleagues expressed sympathy but even they could not reverse the actions of the elected village head, a powerful Brahmin male. Even though most of the allegations had subsided after about a year, Hema Devi was bitterly aware of her limited abilities, when we spoke with her after her sudden downfall.

In the early phases of the then eulogized Swajal Project, my observations of gender coercions, compared to numerous other (internally produced) stories of gender success, were sometimes blamed as anecdotal; nonetheless, recent critiques show observations similar to mine. The most vociferous of these is Govinda's account of the decision of a women's organization in Uttar Pradesh to discontinue with the Swajal Project because of its "time-bound, single issue-oriented project activities . . . which served to sideline Dalit women" in a project which had promised the prioritizing of women's empowerment and equity (2007: 54). Govinda writes eloquently that such experiences (mentioned in her case study) are rare because most Indian NGOs, created and run by "urban middle-class, mostly upper caste social elites" are ill-equipped to grapple with the complex challenges faced by poor and marginalized Dalit women (ibid: 46).

Indeed, class and caste asymmetries significantly influenced the continuing inequity in the Swajal project. The local NGO implementing the project (in Mala) was widely acknowledged for its commitment to community empowerment and environment sustainability. However, like many such NGOs, upper-caste Hindu values identified and justified the "sanctity" of culture in their work (Mukhopadhya 1995; Govinda 2007). Senior management staff and founder members of this NGO were Brahmins and, with the exception of one woman, all male. There were only two Dalit men holding low-ranking field positions and who, far from influencing decisions, were always conspicuously absent in the areas where the other staff lived, ate and worked together. In the case of this specific project (and this particular NGO), rather than attempting to transform gender inequality, women's subordination and the exclusion of Dalits, the program segregated the upper-caste women and Dalits (both men and women) and provided few opportunities to enable a shift in changing the deep-rooted discriminatory attitudes. By ignoring the complexities of gender and caste issues and opting for simplistic interventions, the project, in fact, ensured the replication and perpetuation of the extant unequal social relations. Consequently, the most needy and the most vulnerable (especially the Dalit women) failed to benefit from the new supplies of water that were ostensibly "on paper" designed primarily for them.

Conclusion

The early feminist romanticizing of women's special association with Nature backfired in the development positioning of an assumed innate link between women and water. Women's socially allocated tasks of fetching and managing water for domestic use provided the sector an easy and reliable way to engage women in making projects successful. Reducing inequity by enabling/requiring men to share women's often perilous water tasks, and/or enabling women and men to identify and address deep-rooted socio-cultural disparities, would not have served the overriding efficiency goal of water projects. Indeed, it is for this reason that neoliberal, demand-led projects were far more intent and focused on "engaging women" than the welfare-based supply-driven approaches.

An assumed group of women as "harbingers" of success in drinking water projects is the commonly-held belief of gender success in the domestic water sector and holds true for community, state and new liberal institutional arrangements. While fears around Dalits polluting water sources remain unresolved, official and neoliberal policies have "occasionally" invited Dalit (men) along with upper caste women to engage in community-level participation. When social identities and responsibilities limit engagement or result in ineffectual representation, the victims are readily blamed for their disinterest.

The drinking water sector is also an excellent example of a continued feminization of women's water responsibilities–a coerced obligation for women to compete and compensate for the political, social and economic deficiencies of flawed policies (Molyneux 2007). Transforming entrenched inequities will require advocacy, action and feminist coalitions to collectively dispel currently produced gender myths (Cornwall et al. 2007). Such tasks, however, do not easily go hand-in-hand with planning for project efficiency goals. Consequently, historically entrenched, complex caste–gender disparities, in relation to carrying water home, continue to be ignored and gender remains a contentious concept in this sector, undermining both women's structural disprivileges as well as their unique disparities.

Notes

* Some of the data and information in this chapter relates to an earlier paper by the author–Joshi, D. 2011. "Caste, Gender and the Rhetoric of Reform in India's Drinking Water Sector," *Economic and Political Weekly, Review of Women's Studies*, 46(18): 56–63.
1. The term *domni* is a derogatory colloquial reference to the Dalit women.
2. In common parlance, Tiwari writes of households and not of women in particular.

References

Ahlers, Rhodante and Margreet Zwarteveen. 2009. "The Water Question in Feminism: Water Control and Gender Inequities in a Neo-liberal Era," Gender, Place & Culture, 16(4): 409–26.

Agarwal, Anil. 1981. *Water, Sanitation, Health, for All: Prospects for the International Drinking Water Supply and Sanitation Decade, 1981–90.* London; Washington DC: International Institute for Environment and Development and Earthscan.

———. 1985. "Politics of Environment–II," in *The State of India's Environment 1984–85: The Second Citizens' Report.* New Delhi. Centre for Science and Environment.

Agarwal, Anil and Sunita Narain. 1997. "Dying Wisdom: Rise, Fall and Potential of India's Traditional Water Harvesting Systems," in *The State of India's Environment: The Fourth Citizens' Report.* New Delhi: Centre for Science and Environment.

Agarwal, Bina. 1992. "The Gender and Environment Debate: Lessons from India," Feminist Studies 18(1): 119–58.

———. 1997. "Environmental Action, Gender Equity and Women's Participation," *Development and Change*, 28(1): 1–44.

Cornwall, Andrea, Elizabeth Harrison and Ann Whitehead. 2007. *Feminisms in Development: Contradictions, Contestations and Challenges.* London: Zed Books.

Cullet, Phillipe. 2009. "New Policy Framework for Rural Drinking Water Supply: Swajaldhara Guidelines," *Economic and Political Weekly*, 44(50): 47–54.

Dayal, Rekha, Christine van Wijk and Neelanjana Mukherjee. 2000. *Methodology for Participatory Assessments with Communities, Institutions and Policy Makers: Linking Sustainability with Demand, Gender and Poverty.* Washington DC: Water and Sanitation Program, World Bank.

Ghosh, Gourishankar. 1989. Proceedings of a regional seminar "*Women* and Water–The Family *Hand pump.*" August 29–September 1, Manila.

Gururani, Shubhra. 2002. "Forests of Pleasure and Pain: Gendered Practices of Labor and Livelihood in the forests of the Kumaon Himalayas, India," *Gender, Place & Culture*, 9(3): 229–43.

Jackson Cecile. 1993. "Doing What Comes Naturally? Women and Environment in Development," *World Development*, 21(12): 1947–63.

Joshi, Deepa. 2002. "The Rhetoric and Reality of Gender Issues in the Domestic Water Sector: A Case Study from India." PhD Thesis, University of Southampton, UK.

Joshi, Deepa and Ben Fawcett. 2006. "Water, Hindu Mythology and an Unequal Social Order in India," in Terje Tvedt and Terje Oestigaard (eds), *A History of Water, Vol. 3: The World of Water*, pp. 119–36. London: IB Tauris.

Kabeer, N. 1994. *Reversed Realities: Gender Hierarchies in Development Thought.* London and New York: Verso.

Kapoor, Dip. 2007. "Gendered-Caste Violations and the Cultural Politics of Voice in Rural Orissa, India," *Gender, Place & Culture*, 14(5): 609–16.

Mies, Mari and Vandana Shiva. 1993. *Ecofeminism.* London: Zed Books.

Molyneux, Maxine. 2001. *Women's Movements in International Perspective: Latin America and Beyond.* New York: Palgrave.

———. 2007. "The Chimera of Success: Gender Ennui and the Changed International Policy Environment," in Andrea Cornwall, Elizabeth Harrison and Ann Whitehead (eds), *Feminisms in Development: Contradictions, Contestations and Challenges,* pp. 227–40. London; New York: Zed Books.

Mohanty, Talpade Chandra. 1991. "Under Western Eyes: Feminist Scholarship and Colonial Discourses," in C. Mohanty, A. Russ. and L. Torres (eds), *Third World Women and the Politics of Feminism.* Bloomingdale: Indiana University Press.

Mosse, David. 2004. "Is Good Policy Unimplementable? Reflections on the Ethnography of Aid Policy and Practice," *Development and Change*, 35(4): 639–71.

———. 2007. "Ecology, Uncertainty and Memory: Imagining a Pre-colonial Irrigated Landscape in South India," in Amita Baviskar (ed.), *Waterscapes: The Cultural Politics of a Natural Resource,* pp. 213–47. Nature, Culture, Conservation Series. New Delhi: Permanent Black.

———. 2008. "Epilogue: The Cultural Politics of Water–A Comparative Perspective," *Journal of Southern African Studies*, 34(4): 939–48.

Mukhopdahayay, Maitrayee. 1995. "Gender Relations, Development and Culture", *Gender and Development*, 3(1): 13–18.

O'Reilly, K., S. Halvorson, F. Sultana, and N. Laurie. 2009. "Introduction: Global Perspectives on Gender–Water Geographies," *Gender, Place & Culture*, 16(4): 381–85.

Prokopy, L. 2004. *The Relationship between Participation and Project Success: Evidence from Rural Water Supply Projects in India.* Environment Department papers. Participation series ENV–002, Purdue University, Indiana.

Rawat, Ajay S. and Reetesh Sah (2009). "Traditional Knowledge of Water Management in Kumaon Himalaya", *Indian Knowledge of Traditional Knowledge*, 8(2): 249–54.

Sangameswaran, Priya. 2010. "Rural Drinking Water Reforms in Maharashtra: The Role of Neoliberalism," *Economic and Political Weekly*, 55(4): 62–69.

Shiva, Vandana. 1998. *Staying Alive.* London: Zed Books.

———. 2001. "World Bank, WTO and Corporate Control over Water," *International Socialist Review Aug–Sep 2001,* http://www.thirdworldtraveler.

com/Water/Corp_Control_Water_VShiva.html (accessed February 25, 2013).

Sinha, Subir, Shubhra Gururani and Brian Greenberg. 1997. "The 'New Traditionalist' Discourse of Indian Environmentalism," *Journal of Peasant Studies*, 24(3): 65–99.

Tamang, Sierra. 2002. "The Politics of 'Developing' Nepali Women," in Kanak Mani Dixit and Shastri Ramachandran (eds), *State of Nepal*, pp. 161–75. Kathmandu: Himal Books.

Tiwari, Rakesh. 2006. "Explanations in Resource Inequality: Exploring Scheduled Caste Position in Water Access Structure," *International Journal of Rural Management*, 2(1): 85–106.

World Bank. 1996. "Rural Water Supply Project to Bring Clean Water, Sanitation Services to 1,000 Villages in India," World Bank Press Release 96/30/EAP, http://www.worldbank.org/html/extdr/extme/9644sa.htm (accessed May 24, 2011).

———. 1998. *Water Resources Management Sector Review: Initiating and Sustaining Water Sector Reforms*. Report No 18356-IN 1998. Available at http://www-wds.worldbank.org/external/default/WDSContentServer/ WDSP/IB/1999/06/03/000009265_3980901105844/Rendered/PDF/ multi_page.pdf (accessed March 25, 2013).

3

Demand-driven Approach vis-a-vis Marginalized Communities

A Situation Analysis Based on Participatory Approaches in Rural Water Supply and Sanitation Programs in Sri Lanka

N. I. Wickremasinghe

Inadequate access to safe drinking water, sewerage and sanitation infrastructure services are some of the leading causes of public health problems in rural areas of Sri Lanka. According to the 2001 Demographic and Health Survey, conducted by the Department of Health, only about 40 percent of the villages and small town population have private access to safe water sources. Neither is the sanitation coverage satisfactory. Almost 30 percent of the population in villages and small towns and more than 60 percent of the estate population does not have access to sanitary latrines (Department of Health 2001).

In the historical background of rural water supply in Sri Lanka, many projects have been funded by various organizations. One of the first ones is the Hand Pump Tube Well (HPTW) program, funded by UNICEF during the Water Decade Service (1980–90). This program was initiated in dry zone districts (in the rural areas) with the objective of solving domestic water supply issues. However, the initiative of community-based approach was introduced to the water supply and sanitation sector with the launch of the Community Water Supply and Sanitation Project (CWSSP), funded by the World Bank (WB) in 1990, along with the Ministry of Housing and Urban Development (MHUD). It was implemented in three districts in Sri Lanka namely Badulla, Ratnapura and Matara Districts. This was followed by an Asian Development Bank (ADB) assisted pilot Rural Water Supply and Sanitation Project (RWS&S Project; implemented in the period of 1995 to 1998) with the objective of testing community-based systems and procedures developed under the CWSSP, in preparation for an ADB-funded, large-scale rural water supply project. Subsequently, the ADB-assisted Third Water Supply and Sanitation Project (TWSSP) commenced its implementation in 1999 and adopted a community-based approach towards sector reform and institutional development.

In March 2001, the National Policy of the RWS&S Project was approved by the Cabinet of Ministers. Some of the more vital components of this policy have been summarized as follows:

(*a*) Fresh water is a finite and vulnerable resource–essential to sustain life and the ecosystem–and is a basic human need, which warrants equitable allocation.

(*b*) Water has an economic value and should, therefore, be recognized as an economic good. But the provision of water supply and sanitation services should be a people-centered and demand-driven activity.

(*c*) Provision of water supply and sanitation together with educating people on hygiene should be considered as integral components of all sector projects/programs.

(*d*) Demand-responsive and participatory approaches to service delivery, beneficiaries' contribution to capital costs and full responsibility for operation and maintenance.

(*e*) The role of the government, provincial council and local government authorities should be regulated and facilitated in the implementation of sector activities, while community-based organizations, the private sector and non-governmental organizations (NGOs) should be the providers of services.

(f) Users should be encouraged to own and manage facilities and assets and share the capital investment incurred in creating the facilities.

(g) Users should bear full responsibility for sustainable operation and maintenance of facilities.

(h) Women should play a central role in the decision-making process of the sector.

The two recent major initiatives in the RWS&S sector are: (a) the WB-funded Second Community Water Supply and Sanitation Sector Project, launched in July 2003, and, (b) the ADB-funded Secondary Towns and Rural Community-based Water Supply and Sanitation Projects, which began in 2003. Table 3.1 summarizes the present status of rural water supply and sanitation at country level.

Table 3.1: Country profile: rural water supply projects

	Funding Source	Implementing Agency	Project Name	Districts Covered	Present Status
1.	World Bank	Community Water Supply and Sanitation Unit–Ministry of Urban Development and Water Supply	Community Water Supply and Sanitation Project	Ratnapura Matara Badulla	1st phase completed. Balance activities are on-going as second phase.
			2nd Community Water Supply and Sanitation Project	Hambantota Anuradhapura Gampha Colombo	On-going.
			World Bank-assisted 2nd Rural Water Supply and Sanitation	Kandy Matale Nuwaraeliya Kurunegala Puttlam Ampara	On-going

(contd.)

(Table 3.1 contd.)

Funding Source	Implementing Agency	Project Name	Districts Covered	Present Status
		Project. Provincial Council acts as the implementing agency and CWSSP acts as the executing agency	Trincomali	
2. Asian Development Bank	National Water Supply and Drainage Board.	3rd ADB-assisted Rural Water Supply and Sanitation (Sector) Projects	Anuradhapura Puttlam Kegalla Hambantota Monaragala Kalutara	Project completed in 2006.
		4th ADB-assisted Rural Water Supply and Sanitation Project	Anuradhapura Polonnaruwa Hambantota Batticaloa	On-going

Source: TWSSP 2009.

At present, 19 out of 24 districts in Sri Lanka have been implementing rural water supply programs, according to the National Rural Water Supply and Sanitation Policy, which work towards providing village water supply, small town water supply, sanitation facilities, hygiene education, and protection of water sources while the project implementing mechanism is the same for all. The characteristics included (*a*) a demand-driven and people-centered approach, (*b*) participatory decision-making, (*c*) beneficiary involvement in managing resources, (*d*) community contributions for capital cost recovery, (*e*) strengthening and institutionalization of beneficiary organizations, (*f*) operation maintenance and management at lowest appropriate level, (*g*) sub-project implementation through partner organizations and, (*h*) involvement of local authorities and provincial councils. These were intended to result in triggering a community empowerment process, which is directly linked to the well-being of the society. Thus, even though RWS&S projects are specifically focused on water supply and sanitation, their strategies and policies are linked to total development of the society and well-being of the community.

Issues for Discussion

This section focuses on the concept of demand-driven approach in participatory development. Most of the project guidelines emphasize these concepts in preparation of project policies and in all stages– from planning to implementation/construction and Operations and Maintenance (O&M).

However, a few evaluation reports on project activities indicate that some marginalized groups have been deprived of project benefits because they were neglected or left out; these groups had failed to meet project requirements, such as the payment of membership fees and qualifying fees, participation in the required number of project meetings, community contributions in terms of labor and cash, and payment of water connection fees, among others (Benefit Monitoring and Evaluation Report 2006). For example, for the construction of a common water supply scheme, the project requires 20 percent of the total project cost to be met through community cash contributions. In addition, the project needs free and unskilled labor throughout the construction phase. For the construction of individual facilities, such as privately dug wells, spring boxes and rain water harvesting tanks, the project provides funds already decided by the cost-sharing module. Common hand pump tube wells and dug well projects are funded by the community-based organizations (CBOs) in accordance with the amount mentioned in the cost-sharing module. Even projects for sanitation facilities that provide only SLR 4000 (US$ 40)–fixed by the Department of Health (DoH) at a subsidized rate– for the construction of a toilet, are a serious matter for consideration due to the market rates of construction of toilets. According to present market rates, the construction of a toilet will cost (at least) SLR 12,500–20,000 (US$ 125–200). Furthermore, 50 percent of these costs have to be borne by the beneficiary. However, it may not be an affordable amount by the poor, marginalized and other vulnerable community members at the village level. These guiding principles are applicable to both the programs. When considering the project processes, the key milestone could be identified as the involvement of the community in the contribution of cash and labor.

Therefore, the key objective of this chapter is to identify the main reasons for the dropping/leaving out of marginalized communities from the projects and its effects in achieving the objectives and sector goals along with suggestions to address issues relating to marginalized communities. It has been observed that the demand-driven approach, necessitating community cash contributions, creates limitations in involving marginalized communities in projects for providing benefits. Some community development principles, such as social justice, equitable distribution and empowering poor communities, are important matters for debate as they indirectly and adversely affect project outputs that include well-being of the community and equitable distribution of project benefits to the community.

Before analyzing these issues, it is important to understand the concepts behind participatory approaches. Participation is one of the leading concepts from the development point of view because it is potentially a vehicle for different stakeholders to influence development strategies and interventions. According to Britha Mikkelsan–"Participation is the voluntary contribution by people in projects, and their taking part in decision-making and it is the sensitization of people to increase their readiness and ability to respond to development projects" (2005: 53). It is also a voluntary involvement of people with the self-determination to change. This definition points towards a valid participation initiated and managed by people themselves, which is the goal of democratic processes in connection with empowerment practices and its final destination is human well-being. Robert Chambers puts forward the following three concepts on participatory development activities (quoted in Cernea 199: 515–38):

(*a*) Cosmetic label–donor agencies and governments require participatory approaches while consultants and managers state that they will be used. However, the reality often is that top down and traditional approaches are adopted all along.

(*b*) Co-opting practice–to mobilize local level labor and reduce costs. Communities contribute their time and efforts to self-help projects with some outside assistance.

(*c*) Empowering process–to empower those who are marginalized, excluded and deprived, especially women.

When juxtaposed against these concepts, the demand-driven approaches followed in rural water supply projects clearly fall under the first two practices. Participation has enjoyed growing acceptance as an alternative approach to development. At the empowering stage, participation is a process through which the stakeholders influence and share the control over those development initiative decisions and resources which affect them. For achieving these objectives through participatory development approaches, the World Bank and the ADB prepared guidelines based on demand-driven and people-centered approaches, participatory decision-making, community contribution, cost sharing module, institutionalization of beneficiary organizations and appropriate technology. These participatory concepts are directly related to community empowerment, leading to self-determination and positive attitudinal changes among the community members. This helps enhance the well-being of the community. Community well-being is related to poverty levels; therefore, provision of safe drinking water and sanitation facilities helps to a great extent in mitigating poverty through the reduction of waterborne diseases.

While examining the participatory development approaches followed in rural water supply programs in Sri Lanka (with these three concepts), it quite clearly indicates a trend to get community participation throughout the project process. It also expects to inculcate a sense of ownership of the physical and non-physical components in the respective communities. Project plans envisage the inclusion of marginalized and vulnerable sections as project beneficiaries. However, as discussed earlier, the actual situation reveals quite a different picture as some needy community members are dropped (quite early in the project process) due to their inability to meet some of the requirements due to extreme poverty. This issue has to be addressed urgently if these projects are to benefit the underprivileged segments of society, who are most vulnerable and in dire need of these services.

According to Table 3.1, there are five projects in operation at present in Sri Lanka at the national level, utilizing WB and ADB funding sources. Out of these, the ADB-assisted Third Water Supply and Sanitation Project (TWSSP) was selected for this study. The TWSSP was completed in 2006 and the project evaluation report indicated that the benefits are being enjoyed satisfactorily by the beneficiary community (Benefit Monitoring and Evaluation

Report 2006). In addition, this project was identified by the ADB as a suitable approach which could be applied to rural water supply practices effectively. However, some evaluation reports on TWSSP also highlighted that the dropout rate of marginalized community is an important factor for consideration in project planning and implementations strategies (Independent Audit on Completed Rural Water 2005).

Field observations, beneficiary participatory survey results in Kegalla District, secondary data in various surveys and review of available reports and discussions with selected key informants related to the project were key methods adopted in data collection, analysis and report writing (Qualitative Information Appraisals 2006).

Key Findings

Overview of the TWSSP

The project commenced in 1999 and was implemented in two phases. The implementation of 474 pipe gravity schemes, 300 pumping schemes, 17 extension of pipe schemes, 22,882 shallow dug wells, 3,337 shallow common wells, 1,073 hand pump tube wells, 39 shallow tube wells, 1,099 individual shallow tube wells, and 14,449 rainwater harvesting tanks were carried out under the TWSSP during the two phases (The Third Quarterly Progress Report 2005: 57). In addition, 87,000 sanitation units were completed (ibid.). The completed facilities for villagers have been managed by the respective CBOs, representing the respective communities as per the original plan. In order to achieve the project objectives, a set of guiding policies for planning, designing and implementation of water supply and sanitation schemes were developed similar to the policies according to Robert Chambers (discussed earlier).

The project had a strong implementation mechanism. A Project Management Unit (PMU) was established at the National Water Supply & Drainage Board (NWSDB) to provide management and policy support to district-level Project Implementation Units (PIU), which were established in each selected district. In addition, Divisional-Level Implementation Units (DIU) were also created at each of the Local Authorities (LA) where the sub-projects (selected for implementation) were located. The DIUs conducted their operations in close collaboration with the respective LA, under the guidance of the PIUs. In the functional area of hygiene education and sanitation, the Department of Health is also actively involved whenever the project needs their assistance and guidance.

Partner Organizations (POs) have been selected from district/ provincial-based NGOs as the partners for overall project implementation activities, especially in the context of rural sub-projects–to ensure effective mobilization of communities under the demand-responsive approach and equitable distribution of project benefits to the recipient communities. These POs have been selected through a competitive bidding process and their operational areas depend on the size of the sub-project and its geographical situation.

Community-based organizations (CBOs), established at each and every rural sub-project area, were the key actors at the grass roots level in overall implementation of their sub-projects from planning to O&M. The CBO membership comprised households which use the water supply and sanitation facilities provided under their sub-project. Each CBO has an executive committee and a governing body for all functions that are part of the implementation and operation of their facilities. Office bearers of the CBOs were selected through a democratic election process, by respective memberships at their annual general meeting or at a special meeting in accordance with their constitution. All CBOs are registered as voluntary organizations at present, but by-laws are under preparation to delegate power to them (through respective Las) as the legal bodies for O&M of water supply systems within their purview. Project funds are directly handled by the CBOs with the agreement of PIUs. POs are assisted to empower CBOs and develop links between project authorities and other relevant agencies.

This project also had a strong sector co-ordination mechanism from the divisional to the national level. According to administrative hierarchy, Divisional Level Co-ordination Committees (DLCCs) were under the co-chairmanship of the head of Divisional Level Administration and the LA Chairman. Relevant officials of the government departments of the division, NGOs working in the area on water supply and sanitation, representatives from the CBOs and the project are the general composition of the committee. Issues of overall project implementation are discussed at the monthly forum to seek assistance from the sector institutions, whenever and wherever necessary. Next is the Provincial Co-ordination Committee (PCC), chaired by the Chief Secretary of the relevant province. District and provincial level representatives, of the sector institutions, are gathered under this to review the progress of the project functions every quarter. The National Steering Committee (NSC) is chaired by the Secretary to the Ministry of Water Supply and Drainage (MWSD). All relevant sector partners at the national level are members of the NSC while the NWSDB facilitates secretarial services. Policy issues are the main concern of the committee as a broader prospect of sector development through the lessons learned from project implementation.

Subsidy Ceiling and Community Contribution

According to Table 3.2, the minimum contribution by the community towards the cost of construction of water supply facilities is 20 percent–the entire cost of unskilled labor is expected to be borne by the community. If the required minimum community contribution cannot be met by the unskilled labor alone, then the community will have to collect the balance in cash or materials. The maximum subsidy per household (HH) by the project and the minimum Maximum Percentage Community Contribution for each type of facility is given in Table 3.2. The project has to strictly maintain the subsidy ceiling in the preparation of project estimates during the planning stage. The CBOs are informed of the actual cost increase (beyond the subsidy ceiling of the project), which has to be met from community contributions. Sometimes, it may result in increasing the originally expected community contributions. The subsidy ceilings have been prepared considering specific situations in the Wet Zone and Dry Zone leading to differences in some subsidy ceilings in these areas.

Project Cycle

According to Table 3.3, the project development process in the TWSSP included several main phases within which some milestones should be achieved by the respective communities, due to the fulfillment targets of demand-driven and participatory approaches. The table also indicates that community members, who wish to obtain water and sanitation services, have to make various types of payment on seven occasions. In addition, throughout the project process, community members show their participation by attending CBO meetings and free labor during the construction phase. According to field observations and experiences, most of the community members sign the documents related to sub-project selection right at the beginning but some of the poorer categories drop

Table 3.2: Subsidy ceiling for water supply facility

Facility/ Level of service	Project/Community contribution			
	Dry Zone District		Wet Zone District	
	Maximum Project Contribution per HH in SLR	Maximum Community Contribution per HH %	Maximum Project Contribution per HH in SLR	Maximum Community Contri- bution per HH %
1. Pipe Water Supply– Gravity	9,000 (US$ 90)	20	9,000 (US$ 90)	20
2. Pipe Water Supply– Pumping	13,000(US$ 130)	20	7,000 (US$ 70)	20
3. Shallow well– Individual (New)	8,000 (US$ 80)	50	8,000 (US$ 80)	50
4. Shallow well– Individual (Rehabilitation)	2,000 (US$ 20)	50	2,000 (US$ 20)	50
5. Shallow well– Common New (One per 5 HH)	3,500 (US$ 35)	20	3,500 (US$ 35)	20
6. Shallow well– Common Rehabilitation (One per 5 HH)	500 (US$ 5)	20	500 (US$ 5)	20
7. Rain water harvesting	9,000 (US$ 90)	20	8,500 (US$ 85)	20
8. Tube Well New (One per 15 HH)	6,000 (US$ 60)	20	5,000 (US$ 50)	20

Source: Guideline on cost sharing module, TWSSP 2000.

out when the qualifying fees is collected. A majority of community members drop out during the collection of cash contributions. Some vulnerable families do not like to contribute labor without any payment because labor is their only means of a daily income. Most of the marginalized communities have faced some difficulties in

Table 3.3: Important components of project development process

Project Development Activities	Means of Verification
1. Pre-project activities: • Public awareness campaign • Selection of sub-project	• More than 80 percent of the community members should sign the project request application form and show the willingness to obtain water supply and sanitation facility.
2. Project development phase: • Community Mobilization • Sub-project development	• Collect SLR 100 (US$ 1) from each family as qualifying fee for the water supply facility. • Collect SLR 50 (US$ 0.50)from each family as qualifying fee for sanitation facility. • CBO to submit the project proposal.
3. Village participatory planning: • Investigation stage • Feasibility stage • Final designing stage	• CBO started collection of 20 percent community contribution. • CBO completed collection of community contributions. • Community preparation of lands and digging pits for toilet construction.
4. Construction phase	• Community gives free labor contribution for construction activities. • Community gives free labor and balance expenses (for subsidy) for toilet construction.
5. Operation and Maintenance (O & M) phase	• Community-paid water connection fees and regular water supply. • Paying of monthly water bills to village water supply schemes.

Source: Guideline for sub-project development, TWSSP 2000.

getting water connections to their houses as they have to pay meter and other required charges which they cannot afford. Consequently, they do not get the connections. Some members, who have already paid their qualifying fees, find it difficult to pay the required community contributions due to poverty. Other families preferred low services of water supply options (such as private or common dug well) due to their affordability.

Coverage

According to Table 3.4, participatory impact evaluations in TWSSP revealed that most of the village level administrative divisions known as Grama Niladhari Divisions (GNDs) have water supply and sanitation coverage up to 50 percent. However, many schemes and beneficiaries have not obtained their water supply connections (even after the construction is complete) due to poverty. In this table, average water supply coverage is 50 percent while average sanitation coverage is 25 percent. Remaining percentages represent those community members who are unable to meet the necessary payment or those who have already obtained water supply and sanitation facilities. However, it is assumed that the percentage representing the marginalized communities is 50 percent and 75 percent with respect to water supplies and sanitation, respectively.

The recognition of Samruddhi and Non-Samruddhi recipients– that is, the official poverty line introduced by the Sri Lankan government for providing a subsidy for living in the country– is another measure used in the identification of marginalized

Table 3.4: Water Supply and sanitation coverage in selected sub-projects in Kegalla District

Name of the WS Scheme –Sub- project	No. of families' request WS&S	Data of Wealth Ranking According to Participatory Rural Appraisal (PRA) (%)			Increase Protected Coverage (%)	
	Facilities	Poor	Middle	Rich	Water	Sanitation
1. Mawela	419	54	37	9	54	21
2. Ihala Lenagala	302	84	11	5	53	23
3. Rahala	435	62	31	7	54	20
4. Pitiyegama	209	56	35	9	52	28
5. Kandewatta	346	49	45	6	52	19
6. Mahantegama	257	60	35	5	42	21
7. Algoda	341	69	26	5	52	25
8. Bopitiya	332	47	51	2	49	42

Source: TWSSP 2005.

Table 3.5: Provision of water supply by income category

District	Provision of water supply		Non-provision of water supply	
	Samruddhi (percent)	Non-Samruddhi (percent)	Samruddhi (percent)	Non-Samruddhi (percent)
1. Anuradhapura	45	27	5	23
2. Puttalam	31	26	19	24
3. Kegalla	47	45	3	5
4. Kalutara	36	39	14	11
5. Monaragala	42	44	8	6
6. Hambantota	39	40	11	10
Total	40	37	10	13

Source: TWSSP 2006.

communities. Table 3.5 shows the provision of water supplies according to income categories that are based on wealth ranking exercises carried out by the qualitative information appraisal team in Kegalla district, with regard to relevant sub-projects. The income category-wise dropout rate, related to benefits of water supply and sanitation services, is also evident from the table.

Most importantly, Table 3.5 reveals that only 10 percent of the Samruddhi recipients are not receiving water supplies, therefore, proving the figures in Table 3.4, which indicate that 10 percent of the needy community members cannot afford to meet the payments associated with water supplies.

An analysis of the project cycle and water supply and sanitation alternatives indicates that a family contributes a minimum of 10 to 20 days as free labor and an equal number of days in attending various meetings of the CBO. In addition, the cash contribution per family is in the range of SLR 7,500 to 15,000.00 (US$ 75 to 150) including meter charges. Consequently, the total economic value (of all types of contributions) adds up to SLR 12,500 to 17,500 (US$ 125 to US$ 175) per family. For example, minimum cash and labor contributions for pipe-borne water connection is more than SLR 13,700 (US$ 137) excluding monthly water bills. The cost of gravity water supply connection per household is about SLR 11,900 (US$ 119). Rain water harvesting tank costs more than SLR 14,100 (US$ 141) and a private dug well costs about SLR 19,600 (US$ 196). A sanitation facility costs more than SLR 12,050 (US$ 120). Therefore, according to the poverty analysis of Sri Lanka, the income and expenditure pattern of the marginalized and poverty-ridden sections, is not conducive to contribute to the amounts mentioned to avail the services. In Sri Lanka, which is a tropical country, domestic water can be found elsewhere without much difficulty–even in dry zones or drought-prone areas. The key issue here is water quality and quantity, thereby calling for more awareness and motivational and attitudinal changes in all types of community programs. Since project development cycles and cost-sharing modules have a symbiotic relationship, it is very difficult to change the existing system and address poverty group separately.

Way Forward

There is no doubt that the TWSSP and other RW&S programs in Sri Lanka are applying appropriate methodologies to obtain active community participation throughout the project process and inculcate a sense of ownership in the community. This is a useful strategy to achieve the target of community contribution for construction activities. Therefore, all evaluations of the TWSSP highlighted this strategy and recommended that it be emulated by future community participatory programs. The only issue is the high dropout rate among the marginalized sections–not only depriving this highly vulnerable class of people from sharing the benefits of these projects, but also denying them the opportunity to enhance their living standards and well-being. Based on these, there should be fresh and innovative thinking and new strategies for providing demand-driven water supplies and sanitation services to all those who need and deserve it including the marginalized and vulnerable sections.

Therefore, it is prudent to offer suggestions and recommendations that would lead to the inclusion of these neglected social classes as the beneficiaries of projects which provide these services. Before planning the project, it is important to identify the following two categories of community members and consider applying a separate strategy (perhaps, a subsidy scheme) for them:

(*a*) Those who genuinely cannot afford to meet some payments connected with the services

(*b*) Those who cannot afford to contribute free labor (especially daily wage earners) because it would adversely affect their livelihoods.

When studying the project cycle, it was found that identifying the income categories of the villagers has not been done at the planning stage. It is suggested that once the marginalized groups are identified through wealth ranking exercises, they should be subjected to a further probe by a group of enumerators to identify families who fall under the purview of the previously-mentioned two categories.

During the village participatory planning process, a comprehensive analysis of the advantages of various water supply options should be discussed; the technical, socio-economic and environmental feasibility of these options should be reviewed, keeping the cash contribution (per family) within reach of these marginalized people and devise ways to enable them to get maximum benefits from the project activities. The projects could also induce the richer sections to subsidize the poor by paying an extra cash share for construction activities. Accordingly, higher income families can be motivated to pay the cash value of unskilled labor. Therefore, cash contributions from poor families could be covered with this money to provide more unskilled labor to meet the total required labor contribution.

Another option is to introduce an easy payment scheme to help the poorer sections to pay their cash contributions in installments. At the planning stage, the CBO concerned should suggest strategies and activities to enable these people also to enjoy the benefits of the project at par with the other beneficiaries, which should be incorporated into the plan after due consideration. The CBOs can even arrange credit facilities for them, on easy terms, from rural credit institutions and the national banking system and ensure the recovery of such credit. Furthermore, women's ideas and suggestions at this stage would greatly help in finding ways and means of addressing this problem in an appropriate manner.

However, the most important prerequisite for realizing these recommendations is a drastic change in attitudes towards the marginalized segments, across the project area. As a single community, everyone should be ready to serve and respect each other as equals. This would go a long way in assuring equal distribution of project benefits and, thus, ensure the well-being of the community as a whole. The CBO, the project staff and partner organizations can play a key role in this by taking positive steps to involve the marginalized segments of the society in common social and national events in the village as well as rethink strategies to motivate community members including the marginalized sectors, leading to their empowerment as a whole. It is also very important to revise the constitution of the CBO to incorporate these suggestions. Every effort has to be made to elevate the marginalized segments from their present deplorable status to a respectable level and ensure that they are not left out in the provision of services including the provision of safe drinking water and sanitary facilities.

Acknowledgements

I wish to extend my sincere gratitude to the National Water Supply and Drainage Board for the assistance given to conduct this study. The QIA team deserves thanks for the assistance extended in conducting a participatory impact evaluation of the TWSSP. Finally, a special word of thanks to Mr N. P. Karunadasa, for editing and proofreading this chapter, and to Ms K. K. Dissanayake, for typing it.

References

Mikkelsen, Britha. 2005: *Methods for Development Work and Research: A New Guide for Practitioners.* New Delhi: Sage Publications.

Cernea, M. M. 1995. *Putting People First: Sociological Variables in Rural Development.* Report No. 3.2. Washington DC: World Bank.

Department of Health. 2001: *Health Bulletin: Sri Lanka National Health Statistic Report.* Colombo: Statistic Division.

National Water Supply and Drainage Board (NWSDB). 2000. *Guideline for Sub-Project Development.* Report No. 2.2. ADB-assisted Third Water Supply & Sanitation Project. Colombo: NWSDB.

———. 2001. *National Rural Water Supply Policy.* Report No. 3.2. Draft Final Version of ADB-assisted Third Water Supply and Sanitation Project. Colombo: NWSDB.

———. 2003. *Benefit Monitoring and Evaluation Report.* Final Report No. 5.19. ADB-assisted Third Water Supply & Sanitation Project. Colombo: NWSDB.

———. 2005. *Independent Audit on Completed Rural Water.* Report No. 7.6, Final Report of ADB-assisted Third Water Supply and Sanitation Project. Colombo: NWSDB.

———. 2006: *Benefit Monitoring and Evaluation Report.* Final Report No. 8.12. ADB-assisted Third Water Supply and Sanitation Project. Colombo: NWSDB.

4

IWRM, Well-being and Gender

A Perspective from Bhutan

Gongsar Karma Chhopel

This chapter depicts the scenario of water resources in Bhutan in the context of well-being and gender. Water is, undeniably, one of the most important natural resources and Bhutanese people attribute their Gross National Happiness (GNH) to their virgin, natural environment from where abundant clean water flows. Adoption of the principles of Integrated Water Resources Management (IWRM) and good international practices have been advocated as a model for managing water resources sustainably, efficiently and more holistically. IWRM is being considered as the new paradigm of managing Bhutan's water resources for the present generation as well as for the future. The Water Act of Bhutan, 2011, has the principles of IWRM as its cornerstone, and emphasizes that planning and decision-making on water resources and their development and allocation has to be comprehensively done through the River Basin Management Plan for Drinking, explicitly mentioning that the best water resources shall be made available for drinking purposes, even though sanitation is given foremost priority. Based on the holistic overarching principles of the Water Act, water regulation is being formulated by the National Environment Commission (NEC) (in close consultation with various agencies, sectors and people of 20 districts) to infuse greater details–such as ensuring a certain width of buffer on the river banks, and mandatory minimum ecological flow downstream of the hydropower dams–to honour the sanctity of the riverine ecosystems. Age old traditional practices, based on the principles of equity and justice, will be upheld.

Bhutan remains a nation of happiness amid a natural environment, endowed with plentiful water that has a potential to generate 30,000 MW of hydropower (Water Resources Management Plan 2003). The Bhutanese have always revered Nature, worshipping natural objects and sites, such as wetlands and lakes as abodes of gods and spirits, which has (in turn) helped in the preservation and conservation of watersheds and catchment areas that continue to yield a perpetual flow of abundant water in the streams and rivers of the country. Bhutan's small population, approximately 6,50,000 (Population and Housing Census of Bhutan 2005), is considered a boon as it exerts minimal pressure on its natural resources. The country also has a rich culture and traditions that respect women and elders (in general), parents, and, especially, mothers–that is, the feminine gender. Gender problems are thought to be minimal; nonetheless there are some gender inequities. Women, especially in the rural areas, are in charge of household chores while men take up the outside farm work. Household chores call for more water and indoor plumbing is a rare phenomenon in the rural households. As a result, it can take several hours to fetch water from external resources and, in some cases, from far flung areas. The Public Health Engineering Division of the Ministry of Health, responsible for rural water supply, is in the process of connecting rural communities to water taps–to make water easily accessible to such communes and especially ease the chores of the womenfolk who are responsible for running the household.

Well-being: Gross National Happiness

The well-being of the Bhutanese and their environment could be attributed to the leadership and the visions of successive monarchs to whom the concept of living in harmony with Nature and the well-being of its people has always been a priority. GNH has been an intuitive vision of Bhutanese Monarchs and has been the overarching development philosophy of Bhutan to guide the country's development policies and program. This vision was first expressed by the 4th King Jigme Singye Wangchuck in 1972 and later, in an interview with the *Financial Times of India* in 1986, became a principle to be followed by other countries around the world. The concept of GNH defines Bhutan's development objective as improvement in the happiness and satisfaction of the people rather than the growth of Gross National Product (GNP). GNH states that happiness is the ultimate objective of development. It recognizes that there are many dimensions to development other than those associated with GNP and that development needs to be understood as a process that seeks to maximize happiness, rather than purely economic growth.

Bhutan has identified four major areas as the pillars of GNH–(*a*) economic growth and development, (*b*) preservation and promotion of cultural heritage, (*c*) preservation and sustainable use of the environment, and (*d*) good governance. With the democratization of the Bhutanese governance system, GNH has become vital to develop conscious intellectual thought and ethics and imagination among the Bhutanese. Figure 4.1 illustrates the characteristics of well-being and the suggested means to achieving them:

The long-term prospect for society to reach this goal of responsible well-being will depend on their ability to set limits to their life's projects (such as the number of cars, number of children, and acquisition of wealth, among others) according to the four principles of: (*a*) Livelihood security, (*b*) Sustainability, (*c*) Equity, and (*d*) Capabilities. For example, if a group of people in a village acquired the knowledge and skills to manufacture incense sticks, this might improve their livelihood security. Conversely, other people in the village whose livelihood depends on the collection and sale of incense plants from the forests may find fewer plants to collect. So, while one section prospers, the survival of another is at stake. Over a period, the stock of incense plant may dwindle leading to degradation of the environment and related biodiversity loss. 'Responsible' well-being is about being aware of these possible consequences and taking precautions to mitigate, right from the initial stage.

Water and IWRM

Environment is one of the main pillars of GNH and through its commitment Bhutan has maintained an enviable record in this sphere among many nations globally. The majority of the country's land mass (nearly 72 percent) is forests (National Action Plan on Biodiversity Persistence and Climate Change 2011), while the remaining territory is split up amongst the agriculturally cultivable land, urban areas, rugged mountains, open pastures, and glaciers. More than 50 percent of the total land area is formally protected, guarding some of the richest biodiversity in the world (National Action Plan on Biodiversity Persistence and Climate Change 2011). The United Nations formally recognized the significance of this achievement by awarding one of its 2005 Champions of the Earth awards to His Majesty, the King of Bhutan, and the citizens. Bhutan's commitment to the environment can be attributed to its system of beliefs and values in which environment is placed at the core of development strategy.

Water, a major and critical part of the environment, is also a vital resource for human survival and economic development; as

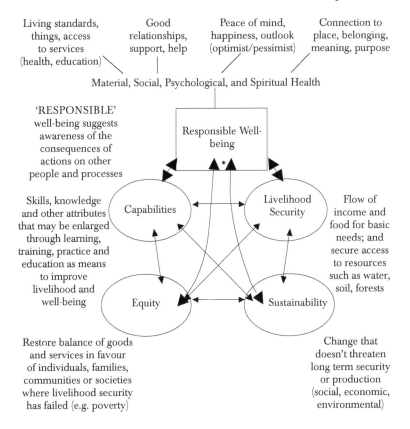

Figure 4.1: Characteristics of well-being

Source: Prepared by the author.

populations and economies grow, the demand for water increases while the availability of the resource remains constant. Bhutan is endowed with rich water resources. The climate of the country is dominated by the monsoon, which sweeps in from the Bay of Bengal during June, becomes intense during July and August and, finally, diminishes during September. The average annual rainfall ranges from nearly 4,000 mm, at altitudes below 500 masl, to less than 500 mm, at altitudes above 4,000 masl (Department of Hydromet Services). 85 to 90 percent of the rainfall takes place in the summer half of the year, that is, from April to September. High rainfall, well-preserved forests and the many glaciers and glacial lakes have given rise to rivers with plenty of water flow in them.

The per capita fresh water availability of the country is among the highest in the world. Studies done by the Department of Energy, with assistance from NorConsult (a Norwegian consulting firm) for preparation of the Water Resources Management Plan, highlight water availability in the following terms:

(*a*) long-term mean annual flow of the entire country is estimated to be 73,000 million m³;

(*b*) per capita mean annual flow availability is estimated at 109,000 m³;

(*c*) per capita minimum annual flow availability is estimated at 20,000 m³.

While at the macro-level water availability is hugely positive, localized water shortages occur in several places due to growing human population, difficult terrain affecting the tapping of water sources, and lack of water storage and poor distribution facilities. A water supply adequacy analysis, carried out in 28 urban centers in 2002–as a part of the preparation of the Water Resources Management Plan–revealed water constraint in 11 towns and predicted water constraint in another seven towns by 2013 (Water Resources Management Plan 2003).

The National Environment Commission (NEC) had conducted baseline water quality survey and monitoring along the major rivers and tributaries between 1997 and 2000. Based on the results, it was concluded that Bhutan's water resources at the macro-level are in a very good state. The data collected through the survey indicates that the main rivers and their major tributaries, with a few exceptions, are of pristine quality. The natural water quality in almost all the rivers and their tributaries is characterized as highly oxygenated, slightly alkaline with low conductivity and no recorded salinity. However, there are localized water pollution problems due to frequent unsanitary conditions along the banks of streams and rivers (Bhutan Environment Outlook 2008).

Water is becoming an increasingly contentious and a sensitive issue in Bhutan, more so due to its necessity for different sectors. Currently, with the installed capacity of 1480 MW, Bhutan earns more than 50 percent of its revenue through the sale of hydroelectricity. With the accelerated development of hydropower to achieve 10,000 MW by 2020, the economy of Bhutan is expected to skyrocket. Given this importance of water in Bhutan, the identification of an appropriate agency to steer the task of managing this precious resource sustainably, efficiently, equitably, and in a more holistic manner has been under discussion since the beginning of 1998. The NEC, owing to its neutral stand with no sectoral biasness, was appointed as the apex body on water in December 2002. Since then, the NEC in collaboration with the Bhutan Water Partnership (BhWP) has successfully developed and completed the Bhutan Water Vision and Bhutan Water Policy–based on which the Water Act of Bhutan, 2011, was formulated by the NEC, in close consultation with stakeholder agencies. The Act views IWRM as the practice of making decisions and taking actions while considering multiple viewpoints of how water could be managed more holistically. These decisions and actions relate to various situations, such as river basin planning, organization of task forces, planning of new capital facilities, controlling reservoir releases, and developing new laws and regulations.

To shoulder these responsibilities, the Government has approved a Water Resources Co-ordination Division under the NEC that will co-ordinate a detailed survey of water resources to establish an inventory for various uses. The quantity and quality of both surface and groundwater will be mapped out. The Division will also be the national focal point to take stock of and guide the future process of water management as well as work on the synergy of professionals from different key sectors. Bhutan also recognizes the critical importance of drinking water and sanitation as a basic pre-condition to lead a healthy and productive life. It is a basic human right and fundamental requirement. This recognition is amply reflected by its constant endeavour to provide safe drinking water and sanitation to all. The Bhutan Water Policy accords the highest priority to water for drinking and sanitation while the Bhutan Water Vision adequately reflects the aspiration and commitment to ensure that water is available in abundance to the people of Bhutan.

Status of Gender Equality and Its Implications for Water Resource Management and Conservation

Gender is generally associated with unequal power and access to choices and resources. The different positions of women and men are influenced by historical, religious, economic and cultural realities. These relations and responsibilities can and do change over time. Water sector decisions have typically excluded women and, at the same time, have followed and practiced sectoral approaches towards development and management of water resources, resulting in differential and adverse impacts on poor women and men, due to ecological and social imbalances (GWA and Cap-Net 2005).

The differences and inequalities between women and men influence how individuals respond to changes in water resources management. Understanding gender roles, relations and inequalities can help explain the choices people make and their different options. It has become increasingly accepted that women play an important role in water management that could be enhanced through a gender mainstreaming strategy. Involving women and men in integrated water resources initiatives can increase both project effectiveness and efficiency (GWA and UNDP 2006). The importance of engaging both women and men in the management of water and sanitation has been recognized at the global level, since the 1977 United Nations Water Conference at Mar del Plata, the International Drinking Water and Sanitation Decade (1981–90) and the International Conference on Water and the Environment in Dublin in January 1992, which explicitly recognizes the central role of women in the provision, management and safeguarding of water. A separate reference is also made to the involvement of women in water management in Agenda 21[1] (Chapter 18) and the Johannesburg Plan of Implementation. Moreover, the resolution establishing the International Decade for Action, "Water for Life"

(2005–15), calls for women's participation and involvement in water-related development efforts. The following water-related activities could reduce women's disadvantage in the sector:

(a) Reduced time, health and care-giving burdens from improved water services give women more time for productive endeavours, adult education, empowerment activities, and leisure activities, among others.
(b) Convenient access to water and sanitation facilities increase privacy and reduce risk of sexual harassment or assault to women and girls, while gathering water.
(c) Higher rates of child survival are a precursor to the demographic transition to lower fertility rates; having fewer children reduces women's household responsibilities and increases their opportunities for personal development (Millennium Project Task Force on Water and Sanitation 2005).

In Bhutan, women (in general) enjoy a relatively greater degree of freedom as their situation is largely influenced by the country's Buddhist traditions and values where men and women are seen as equals and women are represented in most spheres of society. However, there are some gender inequities and gaps, especially in socio-cultural issues such as education, employment and salaries/ wages. At the national level, 69 percent males were literate while only 49 percent of the female population had achieved literacy in 2005. Nonetheless, the combined literacy rate of both men and women is 59 percent (Gender, Water and Sanitation 2006).

The proportion of women employed in the civil service comprises 29 percent (Royal Civil Service Commission 2007). Within environment-related agencies, 26 percent of the staff is female. At the management and executive levels, where the decisions are taken, only 21 percent and 5 percent respectively are females. Furthermore, amongst the total 249,030 employed women, 63 percent are engaged in agriculture alone. These figures indicate that women have lesser access to formal sector jobs and income opportunities than men. It also indicates that men enjoy more access to training and higher education opportunities, from stipend as income source, than women. However, among the upcoming younger generation, more girls are being enrolled in schools; data from all 20 districts show the enrolment ratio of boys to girls (at the lower secondary level) as

23,724: 24,212. At the middle secondary level, the number of boys is 17,141 while the number of girls is 16,554. At the higher secondary level, the number of boys is 9,783 and the number of girls is 9,097 (Statistical Year Book of Bhutan 2010). With this encouraging trend of educating more girls, we expect that the gap and the disparity between males and females will drastically be bridged.

Even though there is no hard statistics, it is widely believed that most women tackle household chores, both in the urban and rural areas, which require more water. Women, especially in the rural areas, are in charge of household chores while men take up the outside farm works. Thus, Bhutanese women shoulder more burden imposed by the water-related activities. Although, women inherit and own land in most parts of the country, globally, there is an emerging debate that preferred inheritance by women holds them to the land and its affairs and negates them from taking up other opportunities. This could be one of the reasons why there is a tendency for men to let the women carry out agricultural work (a mainstay in rural areas) as well as the household duties. This means that in rural areas, where there is a direct interaction with Nature and environment daily, a higher proportion of women and children live to deal only with Nature. This makes them more responsible for the management and use of natural resources. Hence, environmental education programs, micro-environment action programs and gender activities in rural areas can prove to be more beneficial by targeting the women in such areas. In the urban areas too, according to the National Plan of Action for Gender, women (more than men) are involved in household activities, thus making them key players in choosing the types of commodities to use in the household and hence, the type of waste generated. In order to address waste issues, it would be worthwhile to target women's groups for awareness programs on waste management.

However, a major gap is found at the planning stage. One of the most crucial inputs for good planning is information; but at the sector levels, the availability of gender disaggregated data is still limited, more so on environment-related project implementation and impacts. For instance, in 2005, the coverage of households with access to improved water facilities was 84.3 percent, while households with access to improved sanitation facilities stood at 81 percent (Population and Housing Census of Bhutan 2005). It is not possible to assess this situation by gender as the data available does not enable such an analysis.

The Government of Bhutan in 2001 appointed the Department of Planning (now the GNH Secretariat) as the focal point for co-ordinating all gender matters. In 2004, it established the National Commission for Women and Children (NCWC) as an independent body to strengthen the review, formulation, co-ordination, and implementation of government commitments under the Convention on Elimination of Discrimination Against Women (CEDAW) and the Convention on the Rights of the Child (CRC). There are also several NGOs concerned with improving women's socio-economic conditions and promoting their participation in development activities. These include the National Women's Association of Bhutan (NWAB); Respect, Educate and Nurture Women (RENEW); Youth Development Fund (YDF); and Tarayana Foundation. NCWC, the national level co-ordinating body for gender issues, works with Gender Focal Persons[2] appointed with various line agencies and ministries. However, while there exists a Terms of Reference (ToR) for the Gender Focal Persons in these bodies, it is not part of the regular Gender Focal Persons official responsibility. Consequently, gender issues tend to be subsumed over other agendas and issues. Further, most gender-related projects in the country are aimed at addressing typical gender issues, such as domestic violence, women's empowerment, training on business skills and others. Few projects, such as nettle weaving in Langthel under the Tarayana Foundation and manufacturing of handmade paper under the NWAB, include sustainable resource use and have links to environmental conservation. Therefore, mainstreaming gender in environment conservation and management has been initiated with a UN-funded project. It pertains to enabling women's as well as men's concerns and experiences as an integral dimension of the design, implementation, monitoring, and evaluation of policies and programs for environment management. However, to make the GNH principle as truly encompassing gender equity and well-being, following gender concerns need to be considered:

(*a*) integrate gender information in sector statistics and information management as the lack of gender disaggregated data limits analysis of gender issues;

(b) assess the implications of land ownership by women for their social and economic development;

(c) assess gender issues in agriculture as an important aspect for mainstreaming gender, since agriculture as an occupation is directly linked with environment (and water) management;

(d) improve women's literacy particularly at the tertiary level since it is critical for influencing women's access to managerial positions in formal employment; and

(e) promote and facilitate engagement of women in the government, with particular emphasis on engaging women in management functions.

Conclusion

The political and economic architecture of Bhutan is structured around maximizing GNH rather than GDP. Various institutions use the GNH index and a series of instruments of policy to construct policies that promote GNH. The recently enacted Water Act 2011 will ensure that the water resources of the country will be managed in the most holistic manner in line with the principles of IWRM. This precious resource for the country, on which the economy of the nation hinges will be harnessed adopting the internationally accepted best, practices. The citizens have been promised assured access to this resource for their basic needs for drinking and maintaining sanitation and hygiene at affordable cost.

The impacts of climate change are also being observed in the form of erratic precipitation patterns, alarming rate of glacial retreat (20–30 meters per year), and drying out of the sources of water. Through the sub-regional summit in November 2011, the Bhutan Climate summit will bring together the four countries on the eastern slopes of Himalayas, viz., Bangladesh, Nepal, India and Bhutan to agree on the 10 year road map on water, food, biodiversity and Energy as an adaptation measure to the impacts of climate change.

Bhutan continues to honor the constitutional requirement of maintaining 60 percent of the areas under forest cover to continue the status of a net absorber rather than emitter of greenhouse gases. Such conservation effort will also ensure that the watersheds and catchment areas are properly taken care to ensure perpetual flow in the river systems of Bhutan for continued hydropower generation.

The current 23 percent of the population that live below the poverty line (less than US dollar 1 a day) will also be automatically lifted out of the poverty quagmire with the government's decentralization policy and giving employment to 100 percent of the job seekers. Since the abuse of women and children tends to happen in poorer sections of the society, the economic progress and distribution of wealth amongst all Bhutanese through gainful employment and opportunities will potentially address some of the existing gender gaps in the country.

Notes

1. Agenda 21 is a non-binding, voluntarily implemented action plan of the United Nations with regard to sustainable development, a product of the UN Conference on Environment and Development(UNCED) held in Rio de Janeiro, Brazil, in 1992. It is an action agenda for the UN, other multilateral organizations and individual governments around the world that can be executed at the local, national, and global levels. The "21" refers to the 21st century. It has been affirmed and modified at subsequent UN conferences.
2. Gender Focal Person is an official/person whose role is to address gender issues within the mandate of his organization.

References

Cap-Net, GWA. 2005. "Gender and Mainstreaming in Integrated Water Resources Management: Training of Trainers Package." Multimedia CD and booklet.

Department of Energy. Annual Meteorological Data from the Department of Hydromet Services, Thimphu.

———. 2003. *Water Resources Management Plan.* Government of Bhutan.

Gender and Water Alliance (GWA) and UNDP. 2006. "Mainstreaming Gender in Water Management." Available at http://www.wsscc.org/resources/resource–publications/resource-guide-mainstreaming-gender-water management (accessed May 14, 2013).

Gender, Water and Sanitation. 2006. "Interagency Task Force on Gender and Water." Policy Brief.

Lenton, Roberto, Albert M. Wright and Kristen Lewis. 2005. *Health, Dignity and Development: What Will It Take?* UN Millennium Project Task Force on Water and Sanitation: Earthscan.

National Action Plan. 2011. *Biodiversity Persistence and Climate Change.* A report prepared by the National Biodiversity Center in preparation of the Bhutan Climate Summit, November.

National Environmental Commission. 2008. "Bhutan Environment Outlook." Royal Government of Bhutan.

Royal Government of Bhutan. 2005. *Population Housing and Census of Bhutan.* Thimphu.

———. 2007. Royal Civil Service Commission. Available at rcsc.gov.bt (accessed May 14, 2013).

———. 2010. *Statistical Year book of Bhutan.* A periodic publication by the National Statistical Bureau. Thimphu.

PART II

*State, Markets and Civil Society:
Changing Configurations in
Water Management*

5

Changing Configurations around the State in Water Resource Management in Relation to Multi-stakeholders' Participation in South Asia

Possibilities and Challenges

E. R. N. Gunawardena

—

Integrated Water Resources Management (IWRM) has been identified as the way forward to address complex issues associated with water resources development and management and, therefore, is being accepted and promoted globally. Reforms, which were considered as prerequisites for its implementation, have been facilitated by capacity-building along with infrastructural development projects funded by international lending institutions and governments.

IWRM was based on four principles, called Dublin Principles, formulated at the International Conference on Water and Environment (in Dublin) in 1992. These four principles are: (*a*) Fresh water is a finite and vulnerable resource, essential to sustain life, development and environment; (*b*) Water development and management should be based on a participatory approach, involving users, planners and policy makers at all levels; (*c*) Women play a central part in the provision, management and safeguarding of water; and (*d*) Water has an economic value and should be recognized as an economic good (GWP 2000). The participation of stakeholders is identified in the second principle and is also emphasized in Agenda 21 of the Earth Summit held in Rio de Janeiro in 1992. Multi-Stakeholders Processes (MSPs) has become a familiar phrase, in recent times, as a vehicle to promote stakeholder participation in all activities associated with water issues.

MSPs aim to bring together all major stakeholders in a new form of communication and decision-finding (along with decision-making) on a particular issue. They recognize the importance of achieving equity and accountability in communication between stakeholders, involving equitable representation of three or more stakeholder groups and their views. MSPs are based on democratic principles of transparency and participation, cover a wide spectrum of structures and levels of engagement and aim to develop partnerships and strengthened networks between stakeholders. They can also comprise dialogues on policy or evolve into consensus-building, decision-making and implementing practical solutions. The exact nature of any such process will depend on the issues, its objectives and participants, the scope of the project and timelines, among others (for details, see www.earthsummit2002.org/msp/index.html).

Though the Multi-Stakeholders Processes is a relatively new term, participatory processes were quite familiar in the irrigated water sector as early as the 1970s. Participation includes people's involvement in decision-making processes, in implementing programs, their sharing benefits of development programs and their involvement in efforts to evaluate such programs (Wijayaratna 2002). Participatory Irrigation Management (PIM), supported by the international community such as International Water Management Institute (IWMI), the World Bank and the Asian Development Bank, with the approval of the various irrigation agencies in respective countries, was accepted by the society with least resistance. Water Users Associations (WUAs) or Farmer Organizations (FOs) were created and empowered as "formal" institutions through required policies and legislations. However, the same cannot be said about the involvement of "multi-stakeholders" in the water sector, which, sometimes, is highly contested.

Participatory Processes vs the MSPs

It is important to distinguish between participatory and Multi-Stakeholder Processes. According to WageningenUR, the idea of participation has been widely adopted at the local level to make development more effective and sustainable, and to empower people to manage their own development. MSPs takes participation to a higher level by bringing governments, businesses and the civil society together in a process of interaction, dialogue and social learning (for details, http://portals.wi.wur.nl/msp/?page=5171). This is shifting the role of the government and opening new (and important avenues) for the civil society (particularly, WUAs, NGOs and local communities) and the private sector to participate in decision-making processes. To build upon this capacity and performance, governance structures can focus on process orientation, formal and informal institutions, inter-organizational relations and co-ordination, bottom-up management, and the expansion of voluntary exchange and self-governance and market-based mechanisms (SIWI 2010). Therefore, the presence of private sector involvement, market-based mechanisms and participation of more sectoral players appears to be the major difference between the participatory processes and the MSPs but this difference has not been fully observed nor been highlighted in studies which are categorized under the multi-stakeholder research.

Therefore, it is important to look at why PIM was successful, how WUAs or FOs (as NGOs) secured space with line agencies as "formal institutions" and why these NGOs find it difficult to up-scale their participation to a higher level of involvement as anticipated in the MSPs. Since the private sector is considered as one of the major stakeholders in this process of water governance, then why is its entry severely contested? Why is it so difficult to come up with an apex body to co-ordinate different sectoral players in the water management sector? An analysis of these issues will help to suggest a pragmatic governance mechanism, which will be acceptable in the national context of each country in managing its water resources.

Multi-Stakeholder Participation in Water Governance

The six chapters, in Part II of this book, have covered different aspects of water sector reforms in South Asia. Chapter 8 on Small Scale Community Water Supply System (SSCWSS) as an alternative to privatized water supply, shows how marginalized and poor communities in urban areas of Kathmandu have demonstrated remarkable success in conserving and utilizing traditional sources, generating water and managing water distribution systems at an affordable price, with minimum external interference. Chapter Seven presents a case study from Western Nepal, of an initiative made at the local level in implementing a pro-poor water supply and sanitation scheme. The subsidy given according to the economic status has played a crucial role in mainstreaming the poor and marginalized communities in development activities. In addition, the micro-finance program provided by user organizations has helped to increase opportunities of income generation by the poor and very poor households which, in the long run, would help these households to pay the water tariff on a regular basis and, thus, guarantee the sustainability of water supply systems.

Since these case studies describe local initiatives with a single use, that is, drinking water supply/sanitation, problems associated with integration and improvement become irrelevant. The complications associated with integration among different uses, users and up-scaling has been amply demonstrated in Chapter 6 titled "Resource Management at Local Level: 'Platform' Approach for Integration" from Nepal, again. However, the 'Platform' could not take its activities forward due to existing disconnect between sectoral policy, plan and legal provisions, and the lack of clarity of the functions of different stakeholders. The required policy for such integration might not yield expected outcomes unless it is supported by legislative and institutional reforms.

The other two chapters from India fall within the scope of multi-stakeholders' definition according to SIWI (2010) where the private sector and market mechanisms played a major role. Independent Regulatory Authority (IRA) is considered as a major pre-requisite to ensure that all stakeholders, including the State, NGOs and private players, could participate in providing water resources and sanitation services. The Indian state of Maharashtra established an

IRA in 2005 and is considered as one of the pioneering states in water sector reform in the South Asian region. Sectoral reforms are being facilitated and sponsored through World Bank-funded programs and projects. The main agenda of these reforms include universal elements, such as full cost recovery, supply on volumetric basis, privatization, and independent regulation. This chapter argues that there is a significant level of incompatibilities and inadequacies of the reform instruments, which give rise to two distinct trends in the behavior of dominant and weak stakeholders in the sector. Alternatively, it also suggests that the alienation of the weak stakeholders can be explained by the "weak-unfriendly" character of the reform measures.

The study from the state of Karnataka (in India) also shows that the expected benefits from water sector reforms do not reach the beneficiaries. It argues that a paradigm shift in addressing water governance issues could reverse progressive efforts such as PIM, which provided space for participatory decision-making. Basic principles related to different aspects of water sector governance, such as water rights, equity, participation, and sustainability, have been overshadowed by focal attention on pricing for economic efficiency, bulk entitlements and private sector participation to be promoted through the IRA.

When compared with Nepal and India, Sri Lanka still does not have a water policy, which is considered as a prerequisite for water sector reforms. The policy process was severely contested since a majority of "multi-stakeholders" were skeptical about the suggested reforms. This chapter discusses the gradual evolution of water sector institutions with the inclusion of relevant stakeholders, supported by policy and legislations, and argues that a transformation of tested local level institutional arrangements appears to work better than the prescribed, imposed external interventions.

Development of Institutions in the Irrigation Sector

Sri Lanka has been identified as a country with long years of hydraulic civilization. The successes of rulers were measured based on the contribution they made in developing the reservoirs and other infrastructure for irrigated agriculture. The institutional mechanism that existed in the past has been exemplary with detailed rules and regulations, including water rights, taxes and water allocation. Local village institutions were responsible for the operation and management of such systems. Community participation in the management of infrastructure in ancient Sri Lanka was mandatory through state law known as *rajakariya*.

These institutional arrangements had been severely affected during the colonial period with the creation of new institutions. The ownership of land was vested with the authority of the Crown and government departments were created to manage them. For example, the Department of Irrigation was established in 1900 and paid employees were responsible for managing irrigation systems. Even after independence (in 1948), farmers expected the government to provide water to their fields and considered such services as the responsibility of the government. Over the years, the budgetary allocation for operation and maintenance of irrigation systems became a burden on the treasury. The inability to charge irrigation water fees, essentially due to political reasons, has aggravated the problem.

As an alternative to agency-managed systems, the participatory and integrated approaches to irrigation management were experimented in major irrigation schemes in Sri Lanka, commencing from the late 1970s. However, such experiments including special projects in selected major irrigation schemes and the Galoya Development program (PIM was first experimented at the Galoya Irrigation Scheme with a major rehabilitation project) could not be sustained due to the absence of an agency and a working arrangement to back them after the project interventions were over (Samarasinghe and Sumanasekere 2005).

Therefore, the then Ministry of Lands and Land Development, in charge of the management of irrigation systems, established the Irrigation Management Division (IMD) in 1984 to implement the Integrated Management of Irrigation Agricultural Settlements (INMAS) in 35 major irrigation schemes. Under this program, a new breed of professionals were employed at the scheme level to organize farmers to form Field Canal Groups (FCGs) as informal groups at the basic hierarchical level of the canal system and Farmer Organizations (FOs) at Distributary Canal (DC) level as formal organizations. In addition, a Project Management Committee (PMC) was established at the scheme level encompassing farmer representatives, elected by FOs and agency officials who are involved in management of irrigation systems. This arrangement of getting water users to work with state officials in decision-making and system operation has improved the performance of the irrigation system. Having observed its success, the Irrigation Department (ID) and the Mahaweli Authority of Sri Lanka (MASL) adopted similar models for participatory management in major irrigation systems and Mahaweli systems that were not covered by the INMAS. Initially, these programs were implemented without legal recognition until the Agrarian Services Act was amended in 1990 to recognize FOs and the Irrigation Ordinance was amended in 1994 to recognize PMC and the appointment of Project Managers.

In 1989, the ID formally agreed (in principle) to handover the management of systems under the DC to FOs and agreed on a cost-sharing mechanism with the FOs, after the government of Sri Lanka accepted the Joint Management (JM) as a policy. The MASL also adopted a similar strategy on the transfer of irrigation management of downstream systems, subsequently, with no cost sharing mechanism.

Limits of Power Transfer

Few pilot studies attempted to up-scale these FOs to function as independent business units. International Water Management Institute (IWMI), under the Shared Control of Resources (SCORE) project and funded by the USAID, created farmer companies and tried to make them function as independent entities in the Hurulu Wewa Irrigation Scheme, an irrigation settlement in the north central dry zone of Sri Lanka. However, these farmer companies could not be sustained after the end of the project period due to various reasons (Jinasena et al. 2000). Another initiative was undertaken in 1997 to up-scale the responsibility provided to the FOs to manage their systems by two government institutions–that is, the IMD and MASL pilot tested self-management of an irrigation system in the Ridi Bendi Ela Scheme (in the north western part) and Chandraika Wewa, (in southern part), respectively. However, this program received heavy criticism from many quarters towards mid-2004 that resulted in the taking over of the irrigation management functions by the ID in 2005. This coincided with the change of political climate within the ID, with a newly appointed minister who was not sympathetic towards any process leading to loss of state sector control over the irrigated agriculture sector. The Ministry of Irrigation along with the Mahaweli and Rajarata Development decided to revive this program in 2006, which did not succeed, and presently, the FOs continue to function only as a co-ordinating, unit to look after various inputs–supply, marketing and other associated functions–whilst the Ridi Benda Ela irrigation scheme operates like any other irrigation system in the country.

Therefore, this shows that there is a limit to transfer of power. The ID supported and empowered the FOs to operate and manage the system under the DC level, since this arrangement has helped the former to reduce operation and maintenance costs. The involvement of FOs in the JM has also helped the agency staff to co-ordinate activities in irrigation systems and ensure efficient distribution of water. However, the agency does not wish to handover the assets and its management to FOs as they do not want to lose their power, influence and associated benefits, which they have been enjoying eversince their respective agencies were established.

Community Empowerment Process in the Water Supply and Sanitation Sector

Water supply and sanitation in Sri Lanka has never been associated with the irrigated agriculture sector. The National Water Supply and Drainage Board (NWSDB), established in 1984, has been responsible for providing water supply and sanitation services as a main line agency. In addition, municipalities, as local authorities, provide this service within city limits.

In order to cater to village communities, the Community Water Supply and Sanitation Project (CWSSP) was launched in 1993. The project adopted an approach where beneficiaries participated fully in the implementation process. After completion of the water scheme, the beneficiaries took responsibilities of operation and maintenance. These water schemes, constructed with the contribution of the Government of Sri Lanka, the World Bank and the Japanese Bank for International Co-operation (JBIC), are formally vested with the beneficiaries after completion. In recognition of the innovative approach adopted by the CWSSP, it was rated as the "best practice" and "well-managed" project among 200 similar projects funded by the World Bank, across the world. In appreciation of this achievement, the World Bank provided a grant of US$ 38.9 million for the implementation of the 2nd phase of the CWSSP.

Based on the experience gained during the implementation of this project, a national policy on the rural water and sanitation sector was formulated.

Institutional Arrangements for Inter-sectoral Co-ordination at the National Level

The preceding sections have described the role that WUAs and CBOs play in the irrigated and water supply sanitation sectors along with the line agencies. However, there is a necessity to co-ordinate all these sectors at the national level, especially with regard to water allocation. There have been three co-ordinating bodies operating at the national level, namely, (*a*) Water Management Panel (irrigation and hydropower sectors), (*b*) Central Co-ordination Committee in Irrigation Management (CCCIM) and (*c*) the National Water Supply and Sanitation Steering Committee (NWSSSC). Unfortunately, since 2002, CCCIM and NWSSSC have been relegated to a state of neglect mainly due to frequent changes of the ministries involved in the water sector. Therefore, the Water Management Panel (WMP) has been the most important co-ordinating body since its inception in 1985 (See Figure 5.1). The WMP is headed by the Director General of MASL and consists of all Heads of Government Agencies concerned with the management and operation of the Mahaweli Ganga Development Project (MGDP), which is the largest multipurpose irrigation development project ever to be carried out in Sri Lanka.

Institutional Arrangements for Inter-sectoral Co-ordination at the Local Level

The District Co-ordinating Committee (DCC) is the main administrative mechanism that co-ordinates activities at the provincial/district level, which consists of all local members of the Parliament, provincial council, Pradeshiya Sabhas, municipal councils, and urban councils in the district, and all administrative officers, heads of departments of provincial councils and regional/district officers representing line agencies (including officials from ID, MASL and NWSDB). Though District Secretaries participate in the national level WMP Meetings and pre-seasonal WMP Meetings, conflicts can occur when water is diverted between reservoirs. These issues are then resolved at DCC and Divisional Co-ordinating Committee (DvCC) meetings.

The District Environmental Committees (DEC) and the District Agricultural Committees (DAC) provide space for NGO representatives. The DEC deals with environmental-related issues at the district level and meets once a month. In addition to the government representatives, the DEC has provided membership opportunities to leading NGOs serving in the district in the field of environmental activities. The DAC consists of representatives of all agricultural-related agencies as well as the FOs.

To strengthen and up-scale these local arrangements at the national level, a pilot project was initiated under the World Bank-funded "Mahaweli Restructuring and Rehabilitation Project" to form a River Basin Committee at the Kala Oya basin. One of the reasons for its failure was the reluctance of local administrative authorities to share power to strengthen a new institution at the river basin level–highlighting the fact that "prescriptive" institutions, such as "River Basin Committees," promoted globally cannot simply be introduced when there are existing organizational arrangements of a similar nature, for example: the DCC in the Mahaweli case. The strengthening of existing organizations with an additional mandate is a much better option, rather than introducing new institutions by external agencies.

Imposed vs Indigenous Policies on Multi-stakeholder Participation

The ADB-funded Institutional Assessment for Comprehensive Water Resources Management (CWRM) project began in 1992 to develop an overarching policy and law to govern water resources and institute a single "apex body" with the responsibility for co-ordinating water-related activities (Ariyabandu 2008), leading to the establishment of the Water Resources Secretariat (WRS) and the Water Resources Council (WRC), respectively, in 1996 by the Government of Sri Lanka. The WRS was responsible for developing the new water resources policy and institutional arrangement that was passed by the Cabinet of Ministers in 2000. Though it was approved, no follow-up action was taken, since there was serious opposition from NGOs, farmers, media and some political parties (Ibid.).

In the meantime, a government change in 2004 has resulted in the formation of a new ministry called Agriculture, Lands,

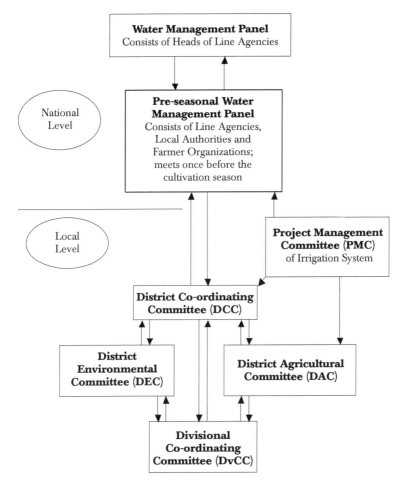

Figure 5.1: Existing co-ordination mechanisms in the
irrigated sector at the local and national levels

Source: Prepared by the author.

Livestock and Irrigation (by amalgamating number of previous
ministries)–which was responsible for water resources utilization,
management and development. The newly appointed minister
was a member of the Janatha Vimukthi Peramuna (JVP) a major
political party of the United People's Freedom Front coalition
government. The JVP is committed to a mixed economic policy

and their political ideology is based on a socialist agenda. This minister had engaged a special committee on June 22, 2004 to develop a water policy, a proposal of which was submitted on August 16, 2004, titled "Deshiya Jala Sampath Pariharanaye, Sanrakshanaya ha Sanvardanaye Moolika Prathipaththi" (that is, "Basic Policies for Utilization, Conservation and Development of Water Resources"). Those who had earlier opposed the ADB-initiated water policy have promoted this indigenous policy since it was formulated without any external influence or intervention.

However, the water policy debate continues with no National Water Resources Policy being available for Sri Lanka. Any public policy becomes useless unless it complies with the constitution of a given country and is accepted by the public. There has not been any attempt during this recent water policy development process to look at the previously-discussed aspects. This policy differs substantially from the ADB-assisted water policy as it totally rejects private sector participation. Water rights including third party rights and transferable water entitlements are supposed to be introduced in the ADB-assisted water policy enabling private sector participation, which facilitate water to be traded in the market as a commodity (Withanage n.d.; Withanage and Tharanganee 2002). This suggestion was also vehemently opposed by the indigenous policy which considers water as a fundamental right. The indigenous water policy proposes to work with existing committees involved with water management at the following four levels, which also differs from what is proposed by the ADB-assisted water policy:

(*a*) National Level
- Central Co-ordination Committee in Irrigation Management (CCCIM)
- National Water supply and Sanitation Steering Committee (NWSSSC)

(*b*) Provincial level
- District Co-ordinating Committee (DCC)
- District Agricultural Committees (DAC)

(*c*) River Basin Level
- Water Panel of the Mahaweli Authority of Sri Lanka

(*d*) Divisional/Project Level
- Project Management Committee

However, there is no representation of the private sector in any of these committees. The new water policy proposes to strengthen the given institutional arrangements with increased participation of the people. Madar Samad (2001) also proposes the same; he says that the progress towards active water markets has been slow and there are a number of serious limitations which suggest that market allocation of water is premature. He concludes that many of the assumed advantages of market allocation can be achieved by concerted efforts at strengthening the institutional framework, which is easier to implement, perhaps less costly, and more readily acceptable by the society. The reasons for the rejection of the ADB-assisted water policy, mentioned in a majority of published literature, were the protests by the civil society, the media and political concerns (Ariyabandu 2008; Nanayakkara 2009).

Foreign-funded NGOs

There are a number of foreign-funded NGOs which are involved with the water and sanitation sectors in Sri Lanka such as Sarvodaya, Plan Sri Lanka, Care International, Intermediate Technology Development Group (ITDG), Cosi Foundation, Netwater, and Neo-Synthesis Research Centre. The first four organizations primarily work towards poverty alleviation. The Cosi Foundation works on sanitation problems whilst Netwater concentrates more on gender issues in the water sector. The Neo-Syntesis Research Centre has been working on environmental conservation. Though water is a cross-cutting theme, in all their activities, these organizations tend to work in isolation without any formal or informal arrangement with line agencies. Some of the work carried out includes watershed management, minor tank rehabilitation, operation and maintenance of minor tanks, ground water development and utilization, rainwater harvesting, crop diversification under minor tanks, institutional linkages, organizational strengthening, farmer and capacity building, rural water supply, health and sanitation, gender and water, and advocacy and policy issues (Pathmarajah and Mowjood 2007). In addition to their parent organizations, international organizations, such as UNDP, World Bank and ADB, provide funding to these NGOs to continue their activities that depends on the support they receive on an ad-hoc basis.

Lessons Learnt and Way Forward

A review of the six chapters (in Part II) provides some generic lessons while indicating directions to be taken in the future in relation to water governance.

(*a*) WUAs in irrigation systems and CBOs in water supply and sanitation sectors have not been evolved within the national context, but catalyzed though foreign-funded projects in the region. The creation of these organizations has facilitated state agencies to carry out their task more conveniently and effectively. Therefore, it is likely that this cohabitation of agency with WUAs and CBOs could be sustained for mutual benefit. However, it is also apparent that state agencies are reluctant to empower non-state sector organizations, beyond a certain limit, and do not want to get rid of their resource base, especially in large and medium irrigation schemes. Handing over water resources and other infrastructure would obviously reduce the influence and importance of such agencies. Therefore, it is unlikely that a total system transfer from the agency to farmer organizations would be a reality since none of the parties want it to happen.

(*b*) Small-scale water supply schemes managed by CBOs, as an alternative to privatized water supply, have demonstrated remarkable success in conserving and utilizing traditional sources, generating water and managing water distribution systems on their own, at affordable prices with minimum external interference. The progressive tariff system formulated by the users' organization has encouraged poor families to pay the tariff on a regular basis. In addition, the micro-finance program, provided by user organizations, has helped to increase the opportunities of income generation by the poor and very poor households which, in the long run, would help these households pay water tariff on a regular basis and, thus, guarantee the sustainability of the water supply system. The integration of activities supporting the livelihoods of people could create an impetus for the inclusion of the very poor households, unlike the mainstream development programs which usually deal only with water supply and sanitation without integrating the activities of livelihood support and limit the opportunities for the very poor households to join the schemes and continue to pay tariff for the operation and maintenance of the system.

(c) In the water supply and sanitation sector, state agencies could concentrate on major population centers whilst the remaining urban and rural population could be covered through water supply projects, operated and managed by the CBOs. This arrangement appears to be the best in view of high capital, O&M and institutional costs required to cover rural areas. Both these systems of line agencies and CBOs working in tandem through Public–Public Partnerships (PUPs), the emerging concept of partnership developed between public water operators, communities and other key groups without a profit motive and on the basis of equality (Corral 2007) appears to be an appropriate model in the South Asian context as well.

(d) The reform process in the water sector to bring efficiency and cost recovery was found to be incompatible and inadequate (on different counts) with the demands put on it by the sector, stakeholders and norms, prescribed by international organizations such as the World Bank. It is also found that in cases like the privatization of irrigation projects, the reform instruments were not compatible with each other. The reform process can only work if the stakeholders, especially the farmers, are equally empowered as the private sector, for example. If not, the weak stakeholders appear to be removed and alienated from the regulatory processes and fail to make any effort or interventions to secure their legitimate rights and benefits. The alienation of weak stakeholders is rooted in their ignorance or lack of awareness about the different decision-making processes such as the privatization of projects or determination of entitlement or tariff as well as about the regulatory mechanisms that are expected to protect their rights during these processes.

(e) This new reform process has created challenges on basic principles of water governance, such as accessibility, affordability, ownership, equity, equitable distribution, delivery, and participation. It demands public participation for democratic decision-making within the water sector.

(*f*) "Public control over governance"–a useful concept that requires people-centered transparency, accountability, participation, and autonomy–also needs to be introduced, with the public having legal rights to intervene at any point of the process when it is felt that the government or implementing/ governing agencies are deviating with results that go against public interest. The ability and the appropriateness of IRA as an institution, to accommodate such basic principles that govern decision-making in the water sector, needs to be studied in detail before subscribing to such an institutional arrangement.

(*g*) The disadvantages of the water reform process (as indicated earlier), over the existing system of water governance of Sri Lanka, was the main reason for the failure of adopting the proposed water policy and water law. Alien administrative structures and procedural requirements prescribed by international agencies might be counterproductive and also could badly impact the gradual development of community-based participatory structures such as the WUAs, as indicated by studies in India. The administrative allocation mechanism is now in place and the present legal mechanism appears to work well in Sri Lanka. The equity, transparency and access to decision-making processes for communities have been assured through the existing arrangements at the local and national levels. Some argue for the introduction of water rights and encourage water transfer through transfer of entitlements for greater efficiency. However, there is a strong objection to this suggestion and opponents argue that the existing system, which appears to work well, should be strengthened. Therefore, the best option would be to strengthen the existing governance mechanism, identify the weaknesses in the present system and address those gradually, rather than impose new structures and regulations which are alien to the local communities.

(*h*) The integration among water users and uses is the biggest challenge for the operationalization of the integrated approach in land and water resource management as envisaged in the water resources policies of many countries in the region. To achieve this, it is necessary to strengthen the relationship among user groups with local institutions, government agencies and external institutions. The policy formulation in itself is inadequate unless it is backed by legislative and institutional reforms.

References

Ariyabandu, R. 2008. "Swings and Roundabouts: A Narrative on Water Policy Development in Sri Lanka." Working Paper 296, Overseas Development Institute, London.

Bandara, K. R. N. 2006. "Development Issues, Options, and Alternatives of Water Resources Management in Kala Oya Basin," in N. D. K. Dayawansa (ed.), *Water Resources Research in Sri Lanka*, pp. 107–16. *Proceedings of the Water Professionals Day*, held on October 1, 2006. Postgraduate Institute of Agriculture, University of Peradeniya, Sri Lanka.

Birch, A. and P. Muthukude. 2000. *Institutional Development and Capacity Building for Integrated Water Resources Management.* Consultancy Report submitted to the Water Resources Council and Secretariat, Ministry of Mahaweli Development, Colombo, Sri Lanka.

Corral, V. 2007. "Public–Public Partnership in the Water Sector," available at Water Dialogues, hppt://www.waterdialogues.org/documents/Public-Private partnershipsintheWaterSector.pdf (accessed January 21, 2011).

Dodds, Felix. 2001. "The Context: Multi-Stakeholder Processes and Global Governance," in Minu Hemmati (ed.), *Multi-Stakeholder Processes for Governance and Sustainability: Beyond Deadlock and Conflict*, pp. 26–38. London; Sterling, VA: Earthscan Publications. Also available online at http://www.earthsummit2002.org/msp/index.html (accessed March 16, 2010).

Global Water Partnership Technical Advisory Committee (TAC). 2000. "Integrated Water Resources Management." TAC Background Paper No. 4, available at http://www.gwptoolbox.org/index.php?option=com_content&view=article&id=36&Itemid=61 (accessed March 6, 2013).

Jinasena, K., D. J. Merry and P. G. Somarattna. 2000. "Institutions for Shared Management of Land and Water on Watersheds." *Proceedings of the National Water Conference on Status and Future Direction of Water Research in Sri Lanka*, November 4–6, 1998. International Water Management Institute, Colombo, Sri Lanka.

Nanayakkara, V. K. 2009. "Perspective on an Overarching Water Policy for Sri Lanka," *Economic Review*, 35(3&5): 6–15.

Pathmarajah, S. and M. I. M. Mowjood. 2007. "Development Initiatives in Water Sector: Lessons learnt by NGOs." *Symposium Proceedings*, November 29, 2006, Cap-Net Lanka, Department of Agricultural Engineering, Faculty of Agriculture, University of Peradeniya, Sri Lanka.

Samad, M. 2001. "Establishing Transferable Water Rights: A Potential Instrument for the Efficiency Gains in Water Resource Allocation?" Seminar on the Draft Water Resources Policy, August 2, 2001, Sri Lanka Association for Advancement of Science (SLAAS), Colombo.

Samarasinghe, S. A. P. and D. U. Sumanasekere. 2005. "Community Empowerment and Management Experiences in the Water Sector." *Proceedings of Consultation on River Basin Management*, September 22, 2005, Lanka Jalani, International Irrigation Management Institute, Colombo.

Wageningen UR. "Multi-stakeholder Processes: Core Concepts," http://portals.wi.wur.nl/msp/?page=5171 (accessed March 15, 2010)

Wijayaratna, C. M. 2002. *Requisites for Organizational Change for Improved Participatory Management*. Report of the APO Seminar on Organizational Change for Participatory Irrigation Management, held in Philippines, October 23–27, 2000. Tokyo: Asian Productivity Organization.

Withanage, H. n.d. *The Dispossession: ADB Water Policy and Privatization: A Case Study in Sri Lanka*. Colombo, Sri Lanka: Environmental Foundation Ltd, Colombo, Sri Lanka.

Withanage, H. and I. Tharaganee. 2002. *Sri Lankan Water Policy: Pricing, Privatizing and Entitlements*. Colombo, Sri Lanka: Environmental Foundation Ltd, Colombo, Sri Lanka.

6

Resource Management at Local Level

"Platform" Approach for Integration

Dhruba Raj Pant and *Khem Raj Sharma*

International Water Management Institute (IWMI)–in partnership with the Department of Irrigation (DoI), Government of Nepal, Stockholm Environment Institute (SEI), University of York and the Institute of Water and Human Resource Development (IWHRD), Nepal–undertook action research on Community-based Integrated Natural Resource Management (CBINRM) from April 2005–November 2008. The action research was initiated to support implementation of the Water Resources Strategy (WRS), 2002, and National Water Plan (NWP), 2005, of the Government of Nepal. The formulation of the WRS was a part of the global phenomenon during the 1990s that emphasized on Integrated Water Resources Management (IWRM) at the basin level. In Nepal, the process was started in 1995, at the behest of the World Bank, with an objective of sustainable use of available water resources with a focus on economic development, environment protection, hazard mitigation, and fostering constructive mechanism for resolving conflicts among and across stakeholders (WECS 2002). The WRS formulation was based on broad stakeholder consultation at the regional and national level through workshops. The Water Resources Strategy (WECS 2002) has identified the following major requirements for water resources planning at the basin level with an integrated approach:

(*a*) Overcoming the overlap of authority and non-harmonization of related acts and regulations; and
(*b*) Mobilization of community level organizations through District Development Committees (DDCs) and Village Development Committees (VDCs).

In the study site, the major stakeholders in resource management were the (*a*) forest and irrigation users group in the upstream, and (*b*) fishers', irrigation and boaters' groups in the downstream besides the international/non-government organizations (I/NGOs). They are formally organized except the Water Users' Group in the upstream. The local institutions–District Development Committee (DDC), Village Development Committee (VDC) and the Municipality– are elected bodies at the district, village and municipality level, respectively. They are politico-administrative units organized on the basis of geographical boundaries and do not exactly match hydrological boundaries, which is important for the integrated approach. Both the users' group and elected institutions do not have institutional linkages or a working relationship among themselves for resource management. This was, first of all, due to prevailing sectoral policies, rules and regulations and the institutions used to plan and implement natural resource management activities. As a result, there were disconnects between various policies on natural resources management and their implementation from the central to the local level. Secondly, the district and village level elected institutions failed to co-ordinate the natural resource management activities at the village and district levels, although they exercise some authority to that effect under the Local Governance Act, 1999. But they use their authority only to collect revenues from the local users, wherever applicable, for example, land tax from irrigation users and local tax from the fishermen.

These institutions, however, did not have elected representation at the time of the action research and were functioning under government-appointed officials because of stalled local elections since 1997. The present set-up, at the national, district and local levels, shows that there are no institutions at the sub-basin level with a mandate to oversee the management of natural resources. The existing politico-administrative units, as mentioned earlier, are hardly capable of co-ordinating and integrating the activities of various resource users and other stakeholders at the sub-basin level that could facilitate the implementation of the WRS (2002) and the National Water Plan (NWP 2005). Therefore, among all the Strategy and the Plan proposals for various institutional reforms, the formation of the sub-basin level committee with representation from concerned stakeholders is the major one. However, both WRS (2002) and NWP (2005) do not say anything about how the sub-basin level committee will be formed.

Objective of the Action Research

The primary objective of the action research was to identify and implement the mechanism for the formation of the sub-basin level committee to help facilitate the implementation of the NWP (2005).

Methodology and Approach of the Study

The methodology applied in this study was, primarily, to review legal, policy and institutional provisions and ascertain if they were conducive or constraining to the integrated approach (Desk-top study in 2005–6), resource availability and stakeholders' access to and control on it and role of various institutions in resource management (Field study in 2006–7), and the formation of the "Platform" approach through stakeholder consultation (Field study in 2007–8). The involvement of the stakeholders in the study and sharing the findings from it constituted an important part in problem analysis and in understanding the need for integrated approach in resource management at the sub-basin. The field study was carried out in the Begnas sub-basin of Kaski District in Western Nepal which has an area of 75.04 sq. km. All six communities (three upstream and three downstream) and 171 households were selected for the study. Based on the household survey carried out under this study in the three communities in the Upper watershed belonging to the Dundkhola catchment (Lamichhane gaon, Thapa gaon and Bhurtel gaon), and the other three in the downstream (Sainik Basti,

Janata ko Chautara and Sat Muhane), 5.4 persons was the average household population.

Expanding the mandate of the existing resource users group and other stakeholders is very important to facilitate the integrated approach. There could be two approaches: (*a*) Creation of sub-basin level committee through external intervention by nominating representation from among users and other stakeholders, and (*b*) Up-scaling the resource users' and other stakeholders' institutional role for the formation of sub-basin level committee. The action research took the second approach in which up-scaling the role of existing institutions through stakeholder consultation had greater prospects for institution building at local levels than creating a new one.

Institutional Dynamics of Resource Use and Users

Overwhelming majority of the people is engaged in agriculture (over 85 percent). The agricultural labor force generally includes the aging population or school-going children as many young, educated people search for off-farm jobs, mainly in the urban centers. The livelihood opportunity from the use of natural resources is limited in the upper catchment as compared to lower catchment, evident from the access to productive land and irrigation (see Table 6.1).

Table 6.1: Percentage of households having cultivated/cropped areas

	Irrigated land (in percent)	Rain fed upland (in percent)	Grassland (in percent)
Upper catchment (n = 86)			
Up to 0.25ha	53	74	50
0.25 to 0.5ha	33	6	5
Above 0.5ha	6	1	1
Valley floor (n = 85)			
Up to 0.25	30	17	20
0.25 to 0.5ha	45	1	5
Above 0.5ha	25	None	None

Source: Field survey 2006.

Very few households have irrigated land exceeding 0.5ha. The percentage of households having rainfed upland is significantly more in the upper watershed. The thatching grass cultivated in the land is not suitable for agriculture. Almost all households in the downstream have some *khet* (irrigated) land, whereas in the upper catchment about 90 percent of the households possess *khet.* The people in the upper watershed are poor in comparison to the downstream users because (*a*) the size of the land holding is small, and (*b*) the poor management of Farmer Managed Irrigation Systems (FMIS), leads to insufficient availability of irrigation water. This is, first of all, due to the traditional land-based water rights practices, which exclude users who do not have customary rights to these irrigation systems. Secondly, the inability of users to mobilize adequate resources internally or externally for the improvement of infrastructure, which could help expand the irrigated area and allow inclusion of new users to provide benefits to the farmers. However, owing to a lack of linkage with external agencies (because of their informal nature), they have not been able to access the external support needed for infrastructure improvement. Nevertheless, the operation and maintenance of the irrigation system by farmers has helped in reducing negative effects downstream. Another important factor is that the upper watershed is the source of water to the Begnas Lake and to the Begnas Irrigation System, constructed by the government, and the drinking water scheme serving the downstream towns.

The Forest Users Group (FUG) has been managing the forest that is community-owned. Forest products that households have access to are agriculture implements, firewood and grass fodder but there is no cash income from these. Organizationally, they are strong but their linkage with other institutions downstream is non-existent. Likewise, there is no organizational linkage between the Community Forest Users Group (CFUG) and Water Users Association (WUA) in terms of managing resources. Due to the community forest in the upstream, the watershed is well-managed and has been effective in controlling the formation of gullies and checking landslides.

Though several water user groups are functioning downstream, the irrigation users group of the government constructed the Begnas Irrigation System and the fishers group, who raise fish in the lake, are prominent users in terms of obtaining benefits from the environmental services of the forest and water management. A few fishers' households, close to the upstream community forest, are also the members of FUG and derive benefits as other members. The water distribution to the branch and tertiary level in Begnas Irrigation System is inequitable, reflecting the users' inability to manage the internal water distribution equally on their own–due to lack of resources and adequate communication among users. Besides, each of the user groups is trying to maximize the benefit from the lake without making substantial contribution for its sustenance. In such a situation, quite often, conflicts arise within and among the water user institutions due to their diverging interests, which is reflected through their organizational undertaking. One of the gaps at the downstream is that an adequate mechanism for common property resource management has not been developed due to lack of defined ownership of the lake.

Theoretically, it is assumed that watershed management in the upper catchment helps in maintaining good environment downstream. The users at the local level were also found to be aware of the interdependence of resources and forest-water (upstream–downstream) linkage, by acknowledging less debris flowing to the lake after the community forestry program upstream. The fishers group has observed changes in water quality, increase in water level and fish production after the plantation and with the increase in crown cover upstream. But, changes that occurred due to the linkage between forest and water are yet to be established through empirical studies. This is fundamental to the integrated resource management and could positively contribute to the establishment of institutional linkages between upstream and downstream users. One of the primary tasks, therefore, is establishing this linkage through the stakeholders' realization that any benefit downstream is due to the action of upstream users and that its cost/benefit needs to be ascertained. Actualization of such linkage at the basin level would be possible only when the governance at the system level is improved. This would facilitate the introduction of appropriate benefit-sharing mechanisms. Foremost, the users need to acknowledge and accept the concept of benefit sharing,[1] which is quite new to them. Inequity in benefit-sharing among and across the users at present is not conducive to promote CBINRM due to lack of stakeholders' interest and participation. In this respect, a holistic approach of catchment management, in consultation with relevant stakeholders, can create a win-win situation for everyone. This is not taking place due to the lack of relevant policies that delineate the roles and responsibilities of both external and local organizations in common property resource management. Besides, the enforcement of any mechanism is not possible without an intermediary that plays the role of mediator between the resource users, upstream and downstream.

Two sets of institutions have a stake in catchment management. They are (*a*) the local level stakeholders, comprising users' organisations, and (*b*) the local level elected institutions. External stakeholders are the government agencies and I/NGOs who influence/control, and facilitate/provide resources and technical support. The institutional linkage among local stakeholders is horizontal whereas their linkage with external stakeholders is vertical and sectoral, with regard to the respective government offices. Both horizontal and vertical linkages are important for functional linkages among the resource users' group and policy, technical, financial, and material support, respectively. Institutionally, the forest users groups and irrigation users groups at upstream are more isolated in terms of their linkage with downstream water users group and the local elected institutions. It is interesting to note that the FMIS at upstream does not have vertical linkage either with government agencies or with the irrigation users downstream. The registration of the WUG, and their constitution, with the DDC or the government agency is required for official recognition.

The National Water Plan (2005) and Sub-basin Committee

The NWP 2005 has emphasized on decentralized integrated resources management by proposing the sub-basin committee (see Figure 6.1) and the District Water Resources Committee (DWRC) which serve as the Water Assembly (water regulating body) at the district level with representation of local level stakeholders. The formation of DWRC is not on the basis of hydrological boundary but on the politico-administrative boundary with representation from local stakeholders and related government agencies at the district level. The plan has also envisaged the Inter-district Water Resources Committee (IWRC) to address the issues of

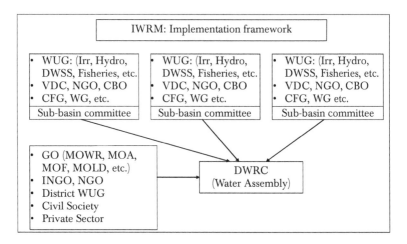

Figure 6.1: Modified version of IWRM implementation framework
for the district proposed in National Water Plan (NWP), 2005

Source: National Water Plan 2005.

hydrological boundary. The DWRC exists at present but its
tasks are limited to the recommendation for issuing license for
hydropower development and registration of water users. The
plan, however, is not explicit in the processes to be followed for the
formation of sub-basin committees and adequate implementation
mechanisms have not been developed, despite major policy focus
of the government.

The "Platform" Approach

The action research was carried out in Kaski District, in the Western
Development Region of Nepal, to facilitate the formation of sub-
basin level committee in the form of "Platform" in 2007.

"Platform" is a venue where resource users and stakeholders are
brought together to discuss issues related to resource management;
it is believed that this process will contribute to the users' and
other stakeholders' understanding and thinking on integrated
resource management (Pant 2011). These types of 'platforms' are
also known as Multi-stakeholder Platforms (MSPs). These MSPs,
depending upon the type and nature of the project, were formed
at different institutional levels for scaling up innovations through

learning alliance approach (Smits et al. 2009); fostering institutional co-operation, communication and public participation for natural resource management (Zeynep 2008; Duncan 2010, Puskur 2010); strengthening small-scale producers' role for increased benefit (Netherlands Development Organization 2010); and organizing users for water resource allocation, distribution and management, especially where basins are closed or closing (Warner 2007).

The "Platform" had representation from members of existing water and forest users group, women groups, local government, NGOs, and advocacy groups and focused on:

(*a*) Institutions: Policy, legal, institutional provisions, role, responsibility, and scope of work of different organizations for natural resource management in the basin context.

(*b*) Linkages: Both vertical and horizontal linkages between water-forest-land-users and linkages between two users' organizations in managing natural resources and overlaps.

(*c*) Issues: Related to policy, legal and institutional provisions, implementation mechanisms; linkage among users, facilitating and constraining factors and their implications in integrated management.

Role of Facilitators

The research team, with the support of the local NGO, worked as external facilitator to promote integrated natural resource management in consultation with the local stakeholders. The representatives, nominated by water users, forest users and other local stakeholders, along with officials from DDC, irrigation, forest office, and Municipality participated in the first "Platform" meeting in 2005 and contributed in the identification of the problem, analyzed gaps in linkages and integrated management of resources, besides

(*a*) preparing and agreeing on functions of the "Platform" followed by the formation of an ad-hoc committee vis-à-vis roles and responsibilities of other organizations;

(*b*) agreeing on areas of work for improving the productivity of resource uses; and

(*c*) preparing and agreeing on an action plan.

The ad-hoc committee prepared the constitution of the "Platform," for recognition as a legal entity, which was approved by the stakeholders' meeting. Finally, the "Platform" was registered in the District Administration Office. Figure 6.2 depicts the approach applied in its formation.

Sub-basin Committee vis-à-vis "Platform" and the Implementation of Action Plan

The sub-basin committee proposed for the implementation of the NWP 2005 has more of a co-ordinating role than being an executing agency implementing activities on its own. The committee has to identify programs for integrated resource management and seek the technical and financial support from the government agencies, NGOs and other development agencies in the district–for the implementation of the activities that help manage resources for the benefit of the local users. Likewise, the district-level agencies are to co-ordinate with the committee when implementing their development activities in that particular sub-basin, so that the activities support the local need and benefit the users. The DWRC is to co-ordinate the activities of both the sub-basin committee and the development agencies through the Water Assembly at the district level (Figure 6.2) for the implementation of the NWP. The full implementation, therefore, demands the intensive interaction between government agencies, sub-basin committees and the DWRC. This was the ideal situation.

However, in absence of the full implementation of the NWP at the time of action research, the DWRC was not fully functional; therefore, linking the "Platform" with the DWRC was not possible. But the "Platform" also required formal recognition to act as a legal entity; therefore, it was officially registered as an NGO with the District Administration Office (DAO), headed by the Chief District Officer (CDO), who is also the chairperson of the DWRC, and has acquired legal recognition. The second meeting of the "Platform" committee in 2007, which was also attended by district-level forest and irrigation officials, local chapter of the Chamber of Commerce and Industry and local dignitaries, recognized the need for preparation of an action plan to keep the "Platform" active. Therefore, the project helped develop action plans for the Begnas catchment management by providing services of a

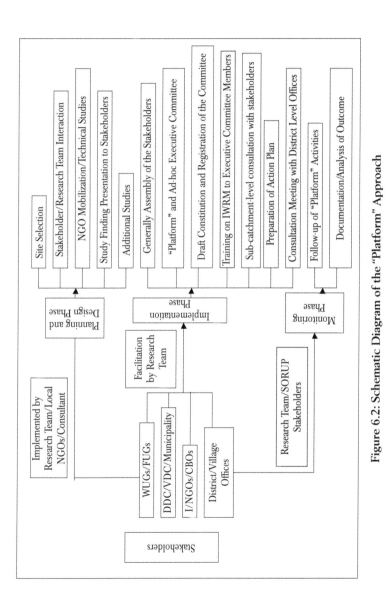

Figure 6.2: Schematic Diagram of the "Platform" Approach

Source: Prepared by the authors.

facilitator. The "Platform" members helped in organizing meeting in various parts of the sub-basin to consult with the stakeholders on activities to be included in the action plan and the modalities for its implementation. The preparation of the action plan was towards creating the linkage between the district-level government and other development agencies and local resource users for operationalizing the NWP 2005. The action plan was expected to be implemented by the "Platform" in co-ordination with government agencies, I/NGOs, users, and other stakeholders, working at the district and local levels, by mobilizing resources from government agencies, locally-elected institutions, I/NGOs, and other public and private organizations. Therefore, the project team organized a district level meeting of the stakeholders to inform them about the formation of the "Platform" and to facilitate linkage between the "Platform" and district level agencies. The latter were supportive of the idea and assured necessary help. However, a follow-up of the "Platform's" activities (after a year of the completion of research work and its operation) revealed that it was not fully functional as was envisaged during its formation. In this period, the "Platform" was only able to organize village-level consultations, executive committee meetings were held regularly and it was exploring suitable modality for its operation. The analysis of the functioning showed the following possible causes for this situation.

First of all, the government has not implemented the NWP 2005 in totality by setting up the required institutional mechanisms at the district and local level–the existing overlaps and contradictions between the various rules, regulation, laws, and institutions have not been removed to facilitate integration. Because of this, the agencies at the district and local level are following the sectoral policies and guidelines, thus, inhibiting the integrated approach at the local level. This is where the technical and financial support from donor agencies (like the World Bank) is required, but its support for institutional reform and capacity building was inadequate in the water sector (Independent Evaluation Group, World Bank 2010). Also, the World Bank has pulled out from its priority to IWRM and, therefore, has made limited progress in the implementation of the NWP 2005 (ibid). This clearly indicates the lack of institutional capability of the respective government to carry on IWRM further. In the absence of necessary reforms at the central level, the district level agencies did not have a clear mandate to follow the implementation of the NWP and have limitations to carry on activities on their own. In this connection, it is also important to take into account that the constitutional assembly, which is formed to draft new constitutions for the country, is expected to come up with a new governance structure for the country, and management of natural resources will be the key component in the new constitution. Therefore, both the political establishment and bureaucracy seem to be waiting for it, without taking further action on the NWP's implementation.

Secondly, the DWRC, which is mandated to play a key role in the basin management, is still dormant and has not been able to provide necessary guidance to the concerned agencies involved in the preparation and implementation of the district plans. Therefore, the "Platform" could not establish the linkage with the DWRC, through which it could have established linkages with other development agencies (government and I/NGOs) at the district level. Consequently, it could not generate necessary support from the district level offices for the implementation of its action plan. It should be noted that the NWP 2005 has not envisaged the functioning of sub-basin committees in isolation.

Thirdly, as per the provision of the NWP, the DDC and VDCs have to mobilize local NGOs and the CBOs. But there is an absence of elected representatives due to stalled local elections for over a decade now. The government officials (who manage these institutions) have little incentive to work with the local CBOs and are more focused on administrative functions. Therefore, the "Platform" members and also the CBOs are reluctant to approach them as easily as they would have approached their elected representatives at district and local level.

Fourthly, the local units of the major political parties have much say in local affairs and most of the decisions of the DDC, VDC and the Municipality are finalized through the committee set-up with their representatives. Thus, the support of these parties is a must for the functioning of any organization/institution at the local level. In that respect, the "Platform" could not garner support from all the parties at the local level due to local political dynamics, as some of the parties thought that they did not have adequate representation in the "Platform" to influence its decision. Therefore, the active support of the local parties was lacking to make it functional.

Finally, the "Platform" tried to initiate some income-generating activities on their own to raise cash resources for their operational expenses and to implement some locally prioritized activities such as lake cleaning. For this purpose, they approached the fishers' group to develop an arrangement for fish-raising in the Begnas Lake. However, the negotiations failed as the fishers group was reluctant to allow a separate group to raise fish in the lake. This could have been a perennial source of income for the functioning of the "Platform." In this connection, it is important to note that the DDC and the Municipality are collecting a tax of NRs 1 per kg of fish from the fishers group without any substantial contribution for the management of the lake.

Overall, it could be said that the lack of clear delineation of the roles and responsibilities of the sub-basin committees and DWRC, vis-à-vis its linkages with other agencies in relation to resource management, seems to be an obstacle to introduce new institutional mechanisms for the implementation of the WRS 2002 and the NWP 2005. Besides, the effort to institutionalize integrated management through the "Platform" could not be sustained due to lack of internal resource generation, the absence of support from the district-level offices, and also because of the phasing out of the project in three years.

Analysis and Conclusions

The action research carried out in one of the catchments in the western part of the country was aimed at operationalizing the integrated approach in resource management, as envisaged in the Water Resources Strategy, 2002 and the National Water Plan, 2005, of the Government of Nepal through the "Platform" approach–which was expected to function as a sub-basin committee facilitating the integrated approach at the sub-basin level by expanding the mandate of the existing institutions. Understanding issues, mainly focusing on (*a*) policy, legal and institutional barriers; (*b*) issues on the dynamics of resource use and users at the local level; and, more importantly, (*c*) exploring the possibility of enhancing the role of existing users' groups in resource management were important components of the study. The desk-top study on the legal, policy and institutional review revealed disconnects and contradictions in sectoral provisions, which were not conducive to the facilitation of the integrated approach at the local level. Therefore, it required timely revisions and amendments in policy and legal provisions for which there is a limitation on the government's capability, requiring external technical and financial support that is lacking.

The action research in the field followed a participatory approach and iterative process. The feedback from users and its incorporation constituted an important aspect of the action research. The intricacies of resource management, the access to and benefit from it, role of various stakeholders, formal/informal institutional linkage among resource users and its implications in management were useful in helping users understand the need for an integrated approach. The interaction between them provided an opportunity to learn from and share their mutual experiences for better resource management. It was felt that the network relationship between various users was not adequately identified and its potential was not fully realized to the benefit of all the stakeholders. This paved the way for the formation of the "Platform" in which the stakeholders, with the facilitation of the research team and the local NGO, actively participated.

The stakeholders were enthusiastic and the initial activities of the "Platform" were encouraging as they drafted their constitution, officially registered as NGOs with the government and prepared a 10 years' action plan and received necessary support from both users and the officials. However, the "Platform" could not take its activities further due to existing disconnects between sectoral policy, plan and legal provisions, clarity of the functions of the sub-basin committee (in the WRS 2002 and the NWP 2005) and mechanisms to link them to the DWRC which, at the time of formation of the "Platform," was still dormant and not in a position to co-ordinate the line agencies' activities and link it to the sub-basin. Therefore, the lack of support from the district level agencies and a lack of elected representatives at the local level left the "Platform" without adequate linkage to the local and district level institutions. This linkage was important in terms of accessing resources, both technical and financial, for the sub-basin planning and implementation of activities for the benefit of the stakeholders. The effort to generate their own resources by the "Platform" was also not successful due to skepticism from some of the resource users and lack of political support from the political parties, both at the local level.

Overall, the lesson learnt from the action research is that any policy formulation in itself is inadequate unless it is backed by legislative and institutional reforms for its effective implementation. The action research, however, was able to fulfill its objectives for the following reasons:

First of all, it developed an approach for organizing stakeholders at the sub-basin level for initiating consultation among themselves on the issues of integrated resources management. The initiation of the consultation in itself was important in the development of an institution for integrated approach.

Secondly, the research was able to make stakeholders understand the intricacies of resource management and linkages of their activities with other resource users by undertaking researches on its various aspects, and access to and control of both resources and the benefits. This enabled the research team to elicit their participation for the "Platform" formation.

Thirdly, the processes applied in the formation of the "Platform" are important in the context of a lack of clear understanding on how to proceed about the formation of the sub-basin committees, as envisaged in the WRS 2002 and the NW 2005, so that these could be replicated in other sub-basins, as well.

Finally, the "Platform" could be up-scaled to function as one of the sub-basin committees when the NWP 2005 comes into full implementation.

Acknowledgements

This research was carried out by International Water Management Institute (IWMI) for the CP-23 Project titled "Linking Community-based Water and Forest Management for Sustainable Livelihoods of the Poor in Fragile Upper Catchments of the Indus-Ganges Basin" of the Challenge Program on Water and Food, part of Consultative Group on International Agricultural Research (CGIAR) and the funding support received under this project is duly acknowledged. The first author was the Project Leader when he was working as Head, IWMI-Nepal from 2001–10. The authors are grateful to the following researchers–Dr Umesh Nath Parajuli, Dr Binod Bhatta, Dr Sabita Thapa, Dr Joe Hill, M. Kamal Gautam, and Ms Pratima Shrestha–who were involved in individual components of the research studies.

Note

1. A system of paying Environmental Services Fee (ESF) to FUGs can be introduced. For example: the fishers group pays NRs 1 per kg to the municipality and the DDC. A similar mechanism could be developed to compensate the FUG upstream.

References

Bandargoda, D. J. 2000. "A Framework for Institutional Analysis for Water Resources Management in a River Basin Context." Working Paper 5, International Water Management Institute (IWMI), Colombo, Sri Lanka.

Duncan, Alan. 2010. "Moving from Project Mode to Innovations Systems Thinking?" Forum for Agricultural Research in Africa, Accra, Ghana. Available at http://www.ilri.org/ilrinews/index.php/archives/1492 (accessed May 2, 2013).

Independent Evaluation Group. 2010. *"Managing Water Resources" in Water and Development: An Evaluation of World Bank Support, 1997–2007, Volume 1, Chapter 3.* NW Washington, DC: International Bank for Reconstruction and Development/World Bank.

Khadka, Shantam S. 1997. "Water Use and Water Rights in Nepal: Legal Perspective," in Rajendra Pradhan, Franz von Benda-Beckmann, Keebet von Benda-Beckmann, H. L. J. Spiertz, Shantam S. Khadka, K. Azharul Haq (eds), *Water Rights, Conflict and Policy: Proceedings of a Workshop Held in Kathmandu, Nepal, January 22–24, 1996*, pp. 13–29. Colombo: IWMI.

Managing Diversity. *Multi-Stakeholder Platforms for Integrated Catchment Management in Water Food Ecosystems–8*, http://www.alterra-research.nl/pls/portal30/www.wageningen-ur (accessed May 2, 2013).

National Water Plan. 2005. Water Energy Commission Secretariat, Government of Nepal. Kathmandu, Nepal. Available at www.wec.gov.np (accessed May 2, 2013).

Netherlands Development Organization. 2010. Available at http://www.snvworld.org (accessed May 2, 2013). Uganda homepage. Agriculture.

Pant, Dhruba, Sabita Thapa, Ashok Singh, Madhusudan Bhattarai, and David Molden. 2002. "Integrated Management of Water, Forest and

Land Resources in Nepal: Opportunities for Improved Livelihood." CA Discussion Paper, IWMI, Colombo, Sri Lanka.

Pant, Dhruba and Pratima Shrestha. 2007. *Institutional Practices for the Management of Natural Resources in Begnas Watershed.* Unpublished report. Kathmandu, Nepal: International Water Management Institute.

Pant, Dhruba. 2011. "Institutional Role and Their Limitations in Integrated Management of Water Forest and Land Resources," in R. N. Gunawardena, Brij Gopal and Hemesiri Kotagama (eds), *Ecosystems and Integrated Water Resources Management in South Asia.* New Delhi: Routledge.

Parajuli, Umesh Nath. 2007. *Assessment of Local Resource Base and its Management in relation to Local Livelihoods in the Begnas–Rupa Basin* Unpublished document.

Puskur, Ranjitha. 2010. "Moving from Project Mode to Innovations Systems Thinking?" ILRI posting 22/02/2010, http://www.ilri.org/ilrinews/index.php/archives/1492 (accessed May 2, 2013).

Samad, M. and D. J. Bandaragoda. 1999. "Methodological Guidelines for Five-country Regional Study on Development of Effective Water Management Institutions." Working Paper, IWMI Colombo, Sri Lanka.

Samad. M. 2001. "Institutional Arrangements for River Basin Management: Some Emerging Issues," in R. N. Kayastha, U. Parajuli, Dhruba Pant, and Chiranjivi Sharma (eds), *Integrated Development and Management of Water Resources: A Case of Indrawati River Basin.* Proceedings of a workshop in Water Energy Commission Secretariat (WECS) and International Water Management Institute (IWMI), Kathmandu, Nepal.

Sharma, Khem Raj amd Binod Bhatta. 2006. "Socio-Economic Study of Six Communities of Begnas Catchment." Unpublished document.

Sharma, Khem Raj, Binod Bhatta and Om Prakash Dev. 2007. "A Review of Policy, Legal and Institutional Provisions for the Management of Natural Resources in Nepal." Unpublished document. Kathmandu, Nepal: Institute of Water and Human Resource Development.

Smits, S., C. Da Silva Wells and A. Evans. 2009. "Strengthening Capacities for Planning of Sanitation and Wastewater Use: Experiences from Two Cities in Bangladesh and Sri Lanka." Occasional Paper Series 44 The Hague, the Netherlands, IRC International Water and Sanitation Centre.

UNECLAC. 1998. "Network for Co-operation in Integrated Water Resource Management for Sustainable Development in Latin America and the Caribbean." Circular nos. 6 and 7, February.

Upadhyaya, Surya Nath. 2007. "National Water Plan and the Legal Regime on Water Resources." Paper presented in the Dialogue Forum on Water Rights, organized by Jalsrot Vikas Sanstha. Kathmandu, Nepal.

Water Resources Strategy. 2002. Water Energy Commission Secretariat,

Government of Nepal. Kathmandu, Nepal. Available at www.wec.gov.
org.np (accessed May 2, 2013).

Warner, J. (ed). 2007. *Multi-stakeholder Platforms for Integrated Water
Management.* Ashgate Studies in Environmental Policy and Practice.
Aldershot, England; Burlington, VT: Ashgate.

Zeynep Sağlam. 2008. "Multi Stakeholder Platforms as a Way towards
Sustainable Water Basin Management: The Case of Istanbul's Ömerli
Basin." MA Thesis, Lund University, Sweden.

7

Accessibility of the Urban Poor to Safe Water Supply

A Case of a Small Town Water Supply Scheme in Nepal

Prakash Gaudel

More than 2.5 billion people, roughly 38 percent of the world's population, lack access to basic sanitation facilities while more than a billion people continue to depend on unsafe drinking water supplies. Much of this population has access to only about 5 liters a day, as opposed to the minimum daily threshold of about 20 liters (UNDP 2006). Worldwide, the proportion of people without improved sanitation decreased by only 8 percent between 1990 and 2006. Without an immediate acceleration in progress, the world will not achieve even half the sanitation target by 2015, which means that the Millennium Development Goals (MDGs), especially on sanitation coverage, are in danger of not being fulfilled (WHO and UNICEF 2008). The focus now needs to be more specific, particularly on poor and marginalized groups where access to basic water supply and sanitation facilities is largely unsatisfactory. The majority of this population is concentrated in underdeveloped countries in Asia, Africa and Latin America. A similar situation is also prevalent in the entire region of South Asia where inequality of human development between men and women is one of the highest, with very strong rural–urban divide in water supply and sanitation services and the access to basic health care facilities. For example, for the year 2009, Nepal had about 80 percent coverage in water supply and 43 percent in basic sanitation facilities, which goes to show that the country is grossly not on track in achieving MDGs for both water supply and sanitation (NPC 2010). In Nepal, around 80 percent of all diseases are attributed to water-and sanitation-related causes and account for around 13,000 child deaths each year from diarrhoeal diseases such as dysentery, jaundice, typhoid, and cholera (MOPPW 2008).

Rapid urbanization and accelerated increase in the urban population is another characteristic of the South Asian region. The coverage for urban water supply is declining as cities are growing while the investment in infrastructure development, to keep pace with the increasing urban water demand, is lagging behind (WECS 2002). Again, the victims are the poor and marginalized people living in urban slums and squatters. Given the widening gap between the richer and poorer sections of the population, the poor and marginalized sections of the society suffer more from the shortages of water (Dasgupta 2003).

Government agencies and development organizations, working in water supply and sanitation sectors in Nepal, are confronted with two pressing situations: (*a*) lower level of coverage of the poor and marginalized sections of the population, both in the rural and urban areas, despite continued investment in the development of water supply and sanitation infrastructure, and (*b*) ensuring sustainability of the completed water supply and sanitation schemes. The public policies generally set favorable environment in pursuing pro-poor development agenda. Water supply and sanitation issues affecting the poor and marginalized communities often tend to be neglected in the course of undertaking the development schemes. The learning has been that the reforms at the policy level alone would be insufficient in increasing the water supply and sanitation coverage of this underprivileged population. The success would critically depend at the level of each development initiative and their abilities to address the specific problems of the poor and disadvantaged groups. Regarding the second concern, the government and development agencies in the region have realized the need for institutionalizing pricing and cost-recovery mechanisms at the level of each scheme to ensure financial viability of the schemes. In fact, in most cases, the financial viability of a scheme is taken as the pre-condition to qualify for external funding. Once again, a uniform approach of pricing and cost-recovery may not be effective in all situations, especially with regard to addressing the payment capacity of these impoverished groups.

Keeping these perspectives in mind, this chapter, based on a case study of a Small Town Water Supply and Sanitation Scheme (STWSS) at Khairenitar Bazar in the western mid-hills of Nepal, demonstrates how micro-policies at the scheme level could be instrumental in increasing water supply and sanitation coverage of the poor and marginalized population. Two micro-policies that supported pro-poor agenda in the scheme were: (*a*) poor households were allowed to pay the water connection fee in installments, and (*b*) differential water tariff based on the volume slab of water consumption. This chapter analyzes to what extent these provisions have been successful in increasing the urban poor's accessibility to safe water and sanitation and to what extent these reforms at the local level have been successful in translating the pro-poor reform agenda into reality.

Locale

The Khairenitar Small Town Water Supply and Sanitation Scheme is located in Khairenitar Bazar of Tanahu district (see Map 7.1), the main commercial area of the Khairenitar Village Development Committee (VDC). The opening of the Prithvi Highway in 1968 (connecting Kathmandu and Pokhara), which passes through Khairenitar Bazar, led to the development of this roadside township primarily due to increased economic opportunities with the opening of the road. People from the adjoining hills started migrating to Khairenitar Bazar after the construction of the road that led to a significant increase in population.

The scheme was designed to serve 808 households in Khairenitar Bazar, in wards nos 8 and 9 of the Khairenitar VDC (ICON-CMS 2003). The construction of the scheme was supported under the Small Town and Sanitation Development Project, implemented by the Department of Water Supply and Sewerage (DWSS), Government of Nepal, under financial assistance of the Asian Development Bank (ADB). The total cost of Khairenitar Water Supply System was NRs 21.06 million (US$ 290,351). The government subsidized 50 percent of the total cost for water supply component while the community was required to contribute the remaining 50 percent including 5 percent cash upfront, 15 percent contribution during the implementation period and 30 percent loan financed through the

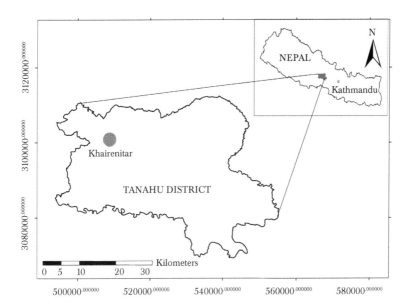

Map 7.1: Map of Tanahu district of Nepal showing Khairenitar

Source: Prepared by the author.

Town Development Fund (TDF) with a payback period of 15 years at an interest rate of 8 percent.

The construction of the scheme included about 6 km of transmission water supply mains, one reservoir of 150 m³ capacity and about 13 km distribution pipelines (ICON-CMS 2003). The scheme taps two spring sources–Jamdi Khola and Jogiko Khet Mul–located within the Khairenitar VDC. This entire system consists of two reservoirs for distribution–a new one with a capacity of 150 m³ and an old one with 100 m³ capacities. Prior to the construction of this scheme, the people in Khairenitar Bazar were served by an older scheme that served water mainly to public taps, though a small number of private connections were also provided to the households.

Methodology

The study involved collection of relevant information from key informants that included project personnel from the DWSS and functionaries of the Khairenitar Small Town Drinking Water Users' and Sanitation Organization. In addition, semi-structured interviews were conducted in a total of 59 very poor households who were beneficiaries of the scheme, to note their perspectives with regard to the processes of inclusion and the benefits that they have been able to realize with the inclusion in the scheme. Out of these households who had already obtained the private tap connection with a subsidy, 43 were from ward no. 8 while the remaining 16 were from ward no. 9 of the project area. Reviews of secondary sources of information were also made to substantiate the primary data collection.

Findings and Discussion

Community Participation

Community participation with the inclusion of all sections of the population is considered essential for any successful community-based intervention (IVO et al. 2008). Realizing that community participation is crucial to the development of the scheme and its subsequent operation, management and upkeep, Khairenitar Water Users' and Sanitation Committee (WUSC) was constituted in 1986–87. The WUSC was formed mainly for the rehabilitation of the old scheme; however, with the scheme qualifying for the assistance under the Small Town Water Supply and Sanitation Development Project, the Khairenitar WUSC was reformed in 2003 and renamed as the Khairenitar Small Town Drinking Water Users' and Sanitation Organization.

At the moment, the main committee of the Khairenitar Small Town Drinking Water and Sanitation Users' Organization has nine functionaries that include the Chairman, the Vice-chairman, the Secretary and Treasurer, and five other members. The functionaries of the executive committee are selected/elected from among the users for three years. Three female members are included in the executive committee as per the provisions of the Small Town Water Supply and Sanitation Sector Project's guidelines. The members meet once a month to decide upon major issues. In addition, three other committees have been constituted to look after specific tasks. An Advisory Committee, consisting of nine members, has been constituted to advise the main committee on important issues. A Monitoring and Supervision Committee, consisting of nine members, has been formed to monitor and supervise the infrastructure and services in the scheme. Similarly, an Accounting Committee, consisting of eight members, has been formed to supervise the accounts and prepare the monthly expenditure report.

Pro-Poor Policies: Subsidies in the Entry Fee

In order to encourage the participation of the poor households and their inclusion in the scheme, the users' organization decided to make subsidy provision in the entry fee. The households making the application for water supply connection were categorized in three categories, based on monthly income, that is, Very Poor, Poor and General. This classification was initially made by the Neighborhood Society Service Centre (NSSC), an NGO, which was involved during the feasibility study and social mobilization phase of the project.

In categorizing the households, those with monthly income less than NRs 3,000 (US$ 42.8) were considered as Very Poor. Initially, out of the total households, 3.4 percent (28 households) were identified in the Very Poor category. But the categorization of the households solely dependent upon the income level was not sufficient to determine their economic status. With this realization, the users' organization incorporated the assets (land/house) along with the income to determine the economic category. Therefore, when any household submits an application for the subsidy in water connection fee, the Monitoring and Supervision Committee determines the economic category of the household with field validation of the information provided by the household regarding its income and other assets. The structure of fees for obtaining the private tap connection for these three categories and provisions for initial deposit and the periods allowed to clear the dues are shown in Table 7.1.

The fee for obtaining private tap connection for the households in the general category was set at NRs 15,000 (US$ 206.73) with an initial deposit of NRs 3,000 (US$ 41.35) and a period of 12 months to clear all the dues, with interest rate of 10 percent. Contrarily, the connection fees for the households in the Poor and Very Poor categories were set to be NRs 12,200 (US$ 168.14) and NRs 7,800 (US$ 107.49) respectively, and households in the Very Poor category were provided a period of 24 months to clear all the dues, after an initial deposit of NRs 1,000 (US$ 13.78). The provision of subsidy is available to only those households who have been living in the project area (prior to its construction)

Table 7.1: Subsidized connection fee structure for
different household categories

S. No.	Economic Category of household	Private-tap Connection Fee (NRs)	Provisions	
			Initial deposit (NRs)	Installment Period
1.	Very poor	7,800	1,000	24 months
2.	Poor	12,200	3,000	12 months
3.	General	15,000 + 10% interest	3,000	12 months

Source: Khairenitar Small Town Drinking Water and Sanitation Users' Organization
2010.

and had obtained a membership in the users' organization. For
new members, who joined the scheme after the construction of
the project, a membership fee of NRs 5,000 (US$ 68.91) and a
connection fee of NRs 15,000 (US$ 206.73) were provisioned by
the users' organization.

The scheme has the capacity of providing private tap connection
to 1300 households. By 2009, 956 connections (including three
community taps) had been distributed which has increased to 1085
by the end of 2010. Of the total connections distributed by this
time, 97 households had obtained the private tap connection on an
installment basis out of which 59 households were from the Very
Poor, 20 households from the Poor and 18 households were from
the General categories, respectively. The remaining households
(mostly from the General category) had obtained the connection
on one-time payment of NRs 15,000 (US$ 206.73) at the time of
the application. In the year 2009–10, the water supply scheme was
extended to Chainpur Area of ward no. 8 of the Bazar. This area
is primarily inhabited by the Dalits (a socially and economically
marginalized group), and other lower caste people. The evaluation
of the economic status and field verification by the Monitoring
and Supervision Committee categorized all 22 households (in
this area) under the Very Poor category, who have obtained the
private tap connection by the end of 2010 at the subsidized rate
of NRs 7,800 (US$ 107.49). The pattern of households of the three
categories obtaining the private water connection, under the subsidy
provisions, over 2006–10 is shown in Fig. 7.1, especially illustrating
the progressively increasing number of households in the Poor and
Very Poor categories.

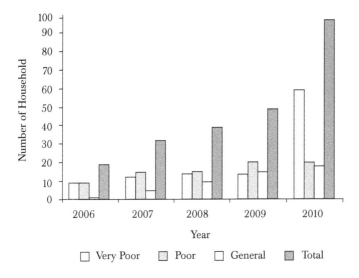

Figure 7.1: Cumulative number of households obtaining subsidized water connection (2006–10).

Source: Khairenitar Small Town Drinking Water and Sanitation Users' Organization (2010).

The users' organization also made provisions for two or more households to give a joint application for sharing a household connection, if they fall under the Very Poor category. These households deposited the amount, sharing the cost of one connection. The provision of initial deposit and the duration for clearing the dues, as applicable to the Very Poor households, were also applicable to those households making an application for a common connection. Three such common tap connections had been distributed by January 2011, of which only two were functional as common taps and one had been converted to a private tap connection. Though this provision made by the users' organization was intended to encourage the inclusion of Very Poor households and reduce their financial burden in obtaining tap connections, it could not produce intended results. The households obtaining a common connection entered into disputes over sharing the water and payment of the monthly water tariff. At present five Very Poor families are benefiting from the two common taps which are functional.

Differential Water Tariff

In small town water supply schemes, the tariff blocks are set on the incremental volume block of water consumption, where the rate for the initial 8–10 m³ is kept at a lower rate, which gradually increases for higher blocks of water consumption (Bhattarai et al. 2007). In Khairenitar, the users' organization has fixed minimum monthly tariff of NRs 120 (US$ 1. 65) for a maximum of 10 m³ of water consumption (see Table 7.2). The old water supply scheme in Khairenitar had a regressive rate of water tariff; therefore, people consuming high volume of water were paying proportionally less than those consuming small volume of water. The households in the "Poor and Very Poor" categories, who were interviewed in the course of the study, expressed satisfaction with the differential water tariff system instituted in the present scheme–as these households have lower level of water consumption. This also creates an incentive for them to consume less water and, accordingly, pay a lower water tariff. According to the users' organization, based upon the tariff collection scenario, about 40 percent of the total consumers use less than 10 units of water per month whereas about 30 percent use more than 20 units per month.

Additional Incentives

(a) Subsidized Latrine: The users' organization came up with a new program as they realized that the Very Poor households were not being able to construct toilets. For the first two years (from start of the water supply scheme), the organization

Table 7.2: Progressive tariff structure
(since April 2010)

S. No.	Units* (Meter Reading)	NRs**
1.	0–10	120
2.	11–15	15 per unit
3.	16–20	18 per unit
4.	21–25	25 per unit
5.	Above 25	30 per unit

Note: * 1 unit meter reading = 1m³ = 1,000 liters; **US$
1 = NRs 72.56
Source: Khairenitar Small Town Drinking Water and
Sanitation Users' Organization 2010.

had made provisions for providing construction materials (free of cost) to assist the construction of toilets for the Very Poor households and improve the sanitation of the area. However, such households could obtain the construction material (including two bags of cement, a toilet pan and two tin-plates as roofing material) only when they obtained a private tap connection. All those Very Poor households who had obtained connections by 2007 availed this facility. This contributed significantly in increasing the number of households with toilets in the project area. By 2009, the total number of households with toilet facilities was 936 that also included households in the Poor and Very Poor categories.

(*b*) Micro-finance Program: The users' organization realized the crucial link between the income level of the households and their abilities of making investments in water supply and sanitation. To encourage the engagement of the Very Poor and Poor households in income-generating activities, the users' organization started a micro-finance program in 2009– whose main objective was to enhance the investment capacity of the poor in water connection and latrine construction. To meet this objective, there are two main areas of investment. The primary area is to help people get loans for a water supply connection and toilet construction through which 15 households have obtained a loan for water supply connection and 14 households for toilet construction. The secondary area is to support income-generating activities. With its support, 19 households are engaged in different income-generating activities. Out of these,nine households are engaged in pig farming, two in poultry farming and the remaining run small grocery shops.

This program is supported by the United Nations Human Settlements Programme (UN-HABITAT), TDF and Poverty Alleviation Fund (PAF). A household cannot obtain a loan directly from the users' organization. UN-HABITAT, under the Water for Asian Cities (WAC) Programme, is supporting the implementation of the water and sanitation related MDGs in Asian cities, specifically promoting pro-poor governance, water demand management, increased attention to environmental sanitation, and income generation for the poor–linked to water supply and sanitation. A household cannot obtain loan directly from the Users' Organization. For this, a group of households (neighborhoods) needs to be formed and registered with users' organization. Each group consists of a minimum of five members. The users' organization provides loan to such group at the annual interest rate of 5 percent. Then, the group provides the loan to the individual household located within its neighborhood. Altogether, 11 such groups are registered to the users' organization but only four are in operation. The number of members varies widely among these 11 groups, from a minimum of five to a maximum of 35 persons per group. Generally, groups comprising larger number of members are not operational due to internal conflicts and lack of co-ordination within the group. The users' organization has already invested NRs 149,000 (US\$ 2053.47) in four neighborhoods (Sundar Tole, Jagriti Tole, Janashristi Tole; and Suryodaya Tole) in the project area. A user can get a loan of maximum NRs 10,000 (US\$ 137.82) from the group at the interest rate of 7–10 percent and can pay it back in installments. The users' organization, through its different groups in the neighborhoods, has adopted a unique approach to provide such loans. The groups in the neighborhood provide loans at lower interest rates (7–8 percent) to those households requesting the loan for obtaining a water supply connection, which is the primary target area of investment of the micro-finance program. Households requesting loans for investing in other income-generating activities, can obtain it at a higher interest rate (9–10 percent), which is the secondary target area of investment. But since these interest rates are lower than those of banks or individual moneylenders available in the market, people are attracted towards this approach of micro-finance program. The interest rates are kept a bit higher in the secondary target area because the rates of return are also higher as compared to those of the primary target area. The flow of money from the users' organization to these groups and, finally, to the individual level has created an environment of secured return of the investment and minimized the administrative cost of the organization, as well. Besides, an individual can also easily get a loan from his/her concerned group of the neighborhood.

To generate financial resources for investment in the micro-finance program, the users' organization negotiated with the donors to reduce the interest rate on TDF loan from 8 percent to 7 percent. This saving (made on the interest) was allocated to create a fund basket that was used to support poor households in starting income-generating activities.

Impacts

More than 70 percent of the population of the area was relying on the existing 165 public taps for water supply, which was unreliable, prior to the new scheme. With its implementation, most of the households are provided with a private connection, which has significantly cut down the time spent by women and the drudgery involved in fetching water from public taps. As per the household survey conducted with 59 Very Poor households, the average time saved (after the project) is estimated to be 68 minutes per day, per household–which is almost 50 percent reduction as compared to the situation before the start of the project. The public taps developed in the old scheme are no longer in use.

Women have been able to use the time saved in starting income-generating and other productive activities such as homestead vegetable gardening and animal husbandry. Out of these 59 households, 11 were involved in such activities. With the availability of water within the homestead, the possibility of starting these activities has been the source of nutrition and additional income for the households. While growing vegetables in homestead gardens has created the incentive of including green vegetables in their daily diet, this has also been an additional source of earning for the Very Poor households as they could sell the surplus vegetables/produce in the local market. Three such households reported an average earning of NRs 3,000 (US$ 41.35) per annum from vegetables production.

According to the users' organization, 98.5 percent of users have toilets in the homestead. Those who do not have toilets are allowed to use their neighbors'. There is no longer open defecation in Khairenitar Bazar and the area has been declared "Open Defecation Free Area." With this declaration, the Khairenitar VDC is also being involved in supporting the Poor households (mainly) in the VDC for constructing the toilets.

Addressing Water Scarcity

With the expansion of private water supply connections and increasing water demand in the area, the Khairenitar Water Supply

Scheme has been facing the problem of water scarcity. The problem is more serious in the dry season (April–June). Therefore, to augment the lean season supply, the users' organization installed a deep tube well/boring system near the Kumble Khola, with an installation cost of about NRs 200, 000 (US$ 2756.34) in 2009. This system pumps the groundwater at the rate of 2 liters/sec. from the depth of 65m and augments water into the main supply system through the old reservoir of 150 m^3 capacity. In 2010, the pump was operated from April to June and the average electricity cost (per month) was around NRs 18,000 (US$ 248.07). The pump is operated only in the dry season but the users' organization has to pay the monthly electricity demand charge of NRs 1,120 (US$ 15.44) throughout the year. Though the installation of the pump has increased the cost of water supply, this has been crucial in mitigating the demand of the users, mainly in the dry season. At present, the water supply system is only able to supply water for six hours a day, in the wet season, and three hours a day, in the dry season. With such rising demand, the users' organization is now in search of new water sources.

Conclusion

This case study of the Small Town Water Supply and Sanitation Scheme in Khairenitar demonstrates the ways and means of including the Poor and Very Poor households among the beneficiaries of the water supply and sanitation scheme. Subsidized connection fee and the water tariff, created by the users' organization, based on the volume slab of consumption and economic status, worked as an incentive for the Poor and Very Poor households to join the scheme. This subsidy has played a crucial role in mainstreaming the poor families and marginalized communities in development activities, who have been regularly paying the tariff. This has been largely possible due to the progressive tariff system formulated by the users' organization.

In order to increase the investment capacity of the Poor and Very Poor households, the users' organization started the micro-finance program that greatly helped to increase income-generating opportunities for them. The users' organization also devised innovative ways of financing the micro-finance scheme through savings on the interest on TDF loans and financial grants from other agencies. The integration of the subsidized latrine construction in the scheme has been yet another incentive for these households to join the scheme.

With increasing coverage of water supply and sanitation in the area, the users' organization has also been able to pay back one-third of the loan by the end of 2010. According to the agreement made with the TDF regarding the loan, the users' organization pays annually a total of NRs 540,844.62 (US$ 7453.76), in two installments. At this rate, the organization will be able to pay back all the loan and interest by the end of 2020.

This study also demonstrated that an integration of activities supporting the livelihood of people could create the impetus for the inclusion of the very poor households. Mainstream development programs, relating to water supply and sanitation, often have a single purpose, which limits the opportunities for the Very Poor households joining the scheme. Though the innovative approach for inclusion in the scheme is exemplary, it would be premature to comment on the sustainability of the scheme as it has been functioning for less than six years; its sustained performance will depend on the abilities of the users' organization in mobilizing enough financial resources to fund the scheme's repair, maintenance and upkeep of the progressively deteriorating physical infrastructure in the days to come.

Acknowledgements

The author would like to express gratitude to Prof. Ashutosh Shukla, Nepal Engineering College, for his valuable guidance and suggestions. The author is also grateful to Mr Sundar Thapa of DWSS, Mr Kamal Dhakal of Bhanubhakta Multiple Campus, Tanahu, and the functionaries of the Khairenitar Drinking Water Users' and Sanitation Organization.

References

Bhattarai, S., B. Shah and S. Aryal. 2007. "Governance on Small Towns Water Supply Project in Nepal." Paper presented at the International Forum on Water Environmental Governance in Asia-Technologies and Institutional Systems for Water Environmental Governance, March 14–15, Institute for Global Environmental Strategies (IGES), Bangkok, Thailand.

Dasgupta, P. 2003. "Valuation of Safe Water Supplies for Urban Households," in K. Chopra, C. H. H. Rao and R. Sengupta (eds.), *Water Resources, Sustainable Livelihoods and Ecosystem Services*, pp. 220–39. New Delhi: Indian Society for Ecological Economics.

Development Research Institute (IVO), Institut de Ciencia i Tecnologia Ambientals (ICTA), International Training Network Centre-Bangladesh University of Engineering and Technology (ITN-BUET), and Central Department of Economics (CEDECON). 2008. "Capacity Building for Enhancing Local Participation in Water Supply and Sanitation Interventions in Poor Urban Areas, Vol. I," http://www.buet.ac.bd/itn/cwp/documents/trainingmaterials/DocumentVolumeI.pdf   (accessed March 17, 2010).

Devkota, D.C. 2007. "Drinking Water Supply and Sanitation Policies and Challenges in Nepal," in *Proceedings of a National Workshop on the "Development of Appropriate Safe Drinking Water Supply System for Slum Dwellers and Squatters in Nepal*, December 17–19, Nepal Engineering College, Kathmandu.

Integrated Consultants Nepal and Consolidated Management Services Nepal (ICON-CMS). 2003. *Feasibility Report of Khairenitar Water Supply and Sanitation Sector Sub-Project.* Unpublished report. Submitted to Department of Water Supply and Sanitation, Kathmandu.

Khairenitar Small Town Drinking Water and Sanitation Users' Organization 2010. Users' Organization Official Records (unpublished).

Ministry of Physical Planning and Works (MOPPW). 2008. *National Urban Water Supply and Sanitation Sector Policy, 3rd Draft, Vol. 3.* Kathmandu: Government of Nepal. Available at http://www.moppw.gov.np/pdf/Urbaneng.pdf (accessed March 21, 2010).

National Planning Commission (NPC). 2010. *Nepal Millennium Development Goals Progress Report.* Kathmandu: Government of Nepal. Available at http://planipolis.iiep.unesco.org/upload/Nepal/Nepal_MDG_2010.pdf (accessed March 18, 2010).

United Nations Development Programme (UNDP). 2006. *Beyond Scarcity: Power, Poverty and the Global Water Crisis.* Human Development Report 2006. New York: UNDP. Available at http://hdr.undp.org/en/media/HDR06-complete.pdf (accessed March 20, 2010).

Water and Energy Commission Secretariat (WECS). 2002. *Water Resources Strategy: Nepal.* Kathmandu: Government of Nepal.

World Health Organization (WHO) and United Nations Children's Fund (UNICEF). 2008. "Progress on Drinking Water and Sanitation: Special Focus on Sanitation," Joint Monitoring Program, New York and Geneva. Available at http://www.unicef.org/media/files/Joint_Monitoring_Report_-_17_July_2008.pdf (accessed March 18, 2010).

8

Small-scale Community Water Supply System as an Alternative to Privatized Water Supply

An Experience from Kathmandu

Hari Krishna Shrestha

—

Traditionally, the people of the Kathmandu Valley were dependent on springs for their daily water needs (Pokhrel 2011). The springs originated at the base of the mountains surrounding the valley. After the development of Kathmandu, Lalitpur and Bhaktapur as urban agglomerations in the Valley, about 2000 years ago, people started constructing dug wells at various locations in and around the settlements to meet the need for water. The water level (in the dug wells) goes deeper and sometimes dries up during long spells of hot, dry periods. During the Kirat regime (800 BCE–300 CE), ponds of various sizes were built in different locations to collect water in the rainy season and to recharge the dug wells during the dry season. The network of rain-fed ponds and dug wells were expanded during the Lichchhavi regime (500–800 CE) by constructing stone spouts connected to the local ponds. The famous stone spouts of Kathmandu (Narayanhiti) and Lalitpur (Mangahhiti) were constructed before the rule of the famous Lichchhavi King, Manadev (Shakya 1994); Narayanhiti was constructed in 464 CE.

With the rise in the population and prosperity of the city states in the 17th and 18th centuries, during the Malla dynasty (1420–1769 CE), springs at the base of the hill slopes were tapped, and using clay pipes and brick canals, the water was brought to the city area from the springs located several kilometers away from the urban areas (see Figure 8.1). For example: the brick canal transported water from the spring source in Tika Bhairav, located 16 kilometers to the south, to the groundwater recharge ponds in Lagankhel, in Lalitpur.

The construction of the stone spouts in the Kathmandu Valley continued till 1828 (Poudel 2003). The outlet points of these drains were decorated with beautifully carved stone spouts, normally located several meters deeper than the surrounding ground level. There are dozens of such stone spouts in all five municipalities of the Kathmandu Valley; the Non-Governmental Organizations Forum for Urban Water and Sanitation (NGOFUWS) estimated (in 2006) that 276 stone spouts in the Valley supply the water need of approximately 10 percent of its population, that is, about 3,00,000 people (UN Habitat 2007; *Gorakhapatra* 2009). In 2006, a survey was conducted by the NGOFUWS on the locations of stone spouts in Lalitpur. These stone spouts discharged water all day long and

Figure 8.1: Schematic of a stone spout in Kathmandu

Source: Shrestha 2009.

were very reliable sources of potable water, until the 1970s. The unused water from the stone spouts was collected in ponds which recharged the dug wells. To maintain the network of springs, canals, stone spouts, and dug wells, local community associations (called *guthi* in Nepali language) were formed by the government and each *guthi* was given the responsibility of cleaning and repairing the specific network. For sustaining the *guthi* system to supply water to the urban areas (through the network), large plots of land were granted to each *guthi*, with a government paper declaring the grant, which was given to the leader of the *guthi* (Shakya 1994). As per the custom at the time, no copies of the land grant were made since the system was based on "good faith" and the ruler retained the right to revoke the grant. The income from the land provided incentive to the members of the *guthi* to regularly clean and maintain the water supply network. In addition, each year on a specific day (called *Sithinakhah* in local language), all water sources and the inlets and outlets of all components of the traditional water supply system (springs, dug wells, stone spouts, pipes, and drains) throughout the Kathmandu Valley are cleaned. The *Sithinakhah* falls just before the onset of the monsoon, as the water tables are their lowest, which eases the cleaning process.

Deterioration of the Traditional Water Supply System

As early as 1875, Kathmandu had a rudimentary piped water supply system (Shakya 1994). The Valley experienced an exponential population growth, after the Rana dynasty in Nepal was overthrown in 1950[1], resulting in the rapid growth of water demand. To deal with the growing water demands, the government built modern water supply systems, by constructing dams and diverting river water to artificial reservoirs and distributing water through pipes. Many people, including several members of the traditional *guthi* system, who depended on stone spouts, started to receive water in their houses through the new government-managed system, which diminished the importance of these spouts and other traditional sources of water. This diminishing status was critical for the proper management of these systems, especially among those who were the age-old users and caretakers of the system. Most of the new users of the stone spouts were recent migrants to the Valley who had no tradition and, hence, no understanding and appreciation of the management of the traditional water supply system.

This disconnect between the new users and the old caretakers of the stone spouts further deteriorated the system since the users had no idea whom to approach if and when the system malfunctioned. The descendants of the original *guthi* members felt less and less responsible for the maintenance of the old system, since they grew up with piped water systems in their house. Many of them converted the ownership of the *guthi* land into their personal ownership and sold the land, thus freeing them from any responsibility for the maintenance of the system. The long neglect of the maintenance and cleaning of the water supply network resulted in either diminished water flow or a complete dry-up of the stone spouts, affecting the water quality. Up to the 1970s, the water from the stone spouts was considered as pure and potable and the local people used it without any treatment. However, after the 1980s, the local people have started to treat the water due to reduction in the quality. Warner et al. (2008) found the water quality of the dug wells and stone spouts of Kathmandu to be very low, with high levels of faecal coliform and nitrate. A study in Bhaktapur also revealed that water from the stone spouts was contaminated with coliform bacteria (Diwakar et al. 2008).

Some of the traditional stone spouts have completely disappeared, as shown in Table 8.1 under the heading "Does not Exist," covered with soil and dirt, while some have been converted into public parks or even private land. A recent survey found that in the Lalitpur Sub-Metropolitan area, seven stone spouts have dried up and another seven have disappeared. Table 8.1 also shows the disappearance of 45 stone spouts from five municipalities of the Valley (data based on NGOFUWS's inventory study conducted in 2006;the number of the stone spouts, which disappeared, is based on interviews with the senior citizens during the study). During an excavation for an underground passage in Bhotahiti, in downtown Kathmandu three decades ago, the remains of a traditional stone spout were found, but instead of resurrecting and putting it into use, the area was covered with soil again; this incidence demonstrates the state of neglect of these traditional systems and the impact of the modern supply system's implementation, without any proper inventory and appreciation of the role of the former.

Table 8.1: Status of stone spouts of Kathmandu Valley

Place	Working with the natural source	Working with City supply line	Does not Work	Does not Exist	Total
Bhaktapur	33	35	18	1	87
Madhyapur Thimi	47	6	9	3	65
Kirtipur	10	0	0	1	11
Lalitpur	47	0	7	7	61
Kathmandu	96	2	34	33	165
Total	233	43	68	45	389

Source: NGOFUWS 2006.

Recent Attempts to Improve Water Supply System in Kathmandu Valley

The rate of growth in water demand (in the Valley), resulting from the rapid population growth (from 3 percent in the 1950s to 5.4 percent in 2000) quickly outpaced the rate of development of the water supply system (Haack and Rafter 2006). By the beginning of the 1980s, the difference between the demand and supply of water became noticeable. In 1988, the Nepal Water Supply Corporation (NWSC) was established to improve the situation by systematic development and distribution of water in the country. The Corporation installed well fields at various locations in the Valley and extracted groundwater to augment the supply from surface water. Partly due to the inefficiency and mismanagement at the NWSC and partly due to the steep rise in water demand, the water distribution situation in Kathmandu went from bad to worse with time. The first reform project supported by the World Bank, from 1991 to 1998, resulted in decreased (rather than increased) access to water in the Valley; the per capita water consumption in liters per day declined from 67 in 1992 to 98 in 1998 (SOE 2001). The Kathmandu Valley's current water demand is about 280 million liters per day (MLD), but the supply is only about 86 MLD during dry season and 105 MLD during wet season and more than one-third of the supply is leaked, resulting in a decline in effective supply. By 2006, only 55 percent of the Valley's population was connected to the NWSC system, and in those households the water flow in the taps was only a few hours a week, even during the monsoon. In the years when the rainfall is low (in the beginning of June, which is the major rice planting season in most parts of Nepal), at different locations in the Valley, the water meant for drinking purposes is diverted to agricultural fields for several weeks, leaving the taps completely dry. In an attempt to reduce the gap between demand and supply, the NWSC started to tap groundwater from the aquifers underneath, at ever higher rates. The current groundwater extraction rate far exceeds its recharge rate, resulting in an unsustainable system–the sustainable groundwater extraction rate is 26.3 MLD; the actual extraction rate is 58.6 MLD. The water table (at various sites) has been decreasing at an alarming rate of 15 to 20 m since 1980 (Khadka 1993), resulting in lower yield from the existing dug wells and stone spouts. Many dug wells have even dried up.

To meet the growing demand of water in the Valley, an inter-basin project was initiated in 1988 to divert water from the Melamchi River (Koshi Basin) to Kathmandu (Bagmati Basin); the completion of the project, however, seems to be a distant target as it was supposed to be completed in 2006. As a result, the pressure on the NWSC to fill the gap between supply and demand of water increased. The different donor agencies, involved in the attempts to solve this problem, found room for improvement in the way water was managed by the NWSC. Up to 35 percent of the water was found to be lost in the system. To reduce the inefficiency by better management of the available resources, under the terms and conditions set for further investment (in the water sector) by the World Bank and the Asian Development Bank, a new semi-private company called Kathmandu Upatyaka Khanepani Limited (KUKL) was established in 2008 to replace the NWSC. However, there does not seem to be any significant improvement in the system so far; the ADB has recently provided KUKL more tankers to distribute water and more generators to pump groundwater during load-shedding hours (Rauniyar 2010). Based on the experiences of the privatization of the water supply system in the urban areas of Asia, Africa and Latin America, Budds and McGranahan noted that "privatization has achieved neither the scale nor benefits anticipated" (2003: 87). With no new sources of water in the system, better management alone is unable to reduce the gap between the supply and demand of water. In 2010, at many places in the Valley, the pipes ran dry for several days a week, with supply of water only for a few hours, resulting in suction (negative pressure) inside the water pipes when the pipes are dry. At many locations in Kathmandu, water and sewer pipes are laid parallel, and next to each other; therefore, during times of suction in water pipes, if the adjacent sewer pipes are leaking, there are high chances of contamination of the water lines. The KUKL has limited capacity to treat the water; only four (out of 14) water treatment plants are in working condition (*Nepal News* 2010); hence, users are forced to either use the untreated water or spend extra resources for water treatment.

Small-scale Community Water Supply System

Instead of waiting for the government to supply more water and alleviate the water supply problem (unlikely to happen anytime soon), local communities, in various locations of the Valley, are coming together and forming community water users' associations to manage the locally available water, in better ways. Since *guthis* are now defunct, new groups have been created to collect and distribute water, most of whom rely on dug wells or stone spouts for the supply; at some places, the locally available water is augmented by piped water supply whenever the former is supplied from the system. A regular fund is collected from the member households[2] or participating families, water collection tanks (of sizes ranging from 500 to 10000 liters) are purchased or are obtained from the government agency, water is collected in these tanks from the dug wells or stone spouts and distributed or sold to the member households, as per the schedule. The community members or managers of the water users' association are responsible for the cleaning of the source areas and maintenance of the system, thereby effectively recreating the traditional *guthi*, except for the land grant. Each of these Small Scale Community Water Supply Systems (SSCWSS) serves 200 to 800 households. Non-members are also allowed to purchase water at low prices. The regular price of 20 liters of potable water in Kathmandu, sold in plastic jars, ranges from NRs 60 to 80 (US$ 1); the same amount of water costs between NRs 3 and 6 at these SSCWSS. A preliminary survey of some of these SSCWSS shows that members of this system are satisfied with the supply of water as each member knows the amount of water available and their fair share of water. At some of the SSCWSS, water is filtered and chlorinated before distribution. Any conflicts among the members are resolved internally. The success stories of these SSCWSS have resulted in the formation of similar systems elsewhere in the Valley; the latest system was created in Tokha (of Kathmandu) in November 2009 (TKP 2009). This chapter discusses the status and management systems of some of these SSCWSS based on the information gathered during the author's field visits, conducted in the period between the months of January and May, 2010.

Management of SSCWSS

The sources of water at the various SSCWSS in Kathmandu are either dug wells or stone spouts and the management of either water distribution system does not differ much. As part of its documentation and conservation activities of historical monuments of Kathmandu, the Patan Conservation and Development Programme–of the Urban Development through Local Efforts (UDLE) of the German Technical Assistance (GTZ), in early 1990s–initiated a system to store water, flowing through the stone spouts at night, and distribute it to the local inhabitants of Alkvahiti (Theophile and Joshi 1992). The Programme found that the combined discharge from seven stone spouts in the northern part of Lalitpur was 1,700,000 liters per day (ibid.), which was enough to serve a large section of the local people. The residents living around other stone spouts have learned from the success of the Alkvahiti experience and adapted the system to suit their local requirements. Later on, the residents living in the vicinity of dug wells also adapted the system.

Since each SSCWSS is small and independent, the management of the water distribution system is flexible. The schedule and duration of the water distribution varies depending on water availability. At some places, each member family gets up to 40 liters of water, per day, free of charge. The financial management and conflict resolution systems vary from one SSCWSS to the next. However, some generalizations can be made regarding the water distribution system at the SSCWSS.

In these systems, depending on the stone spouts, water is collected only at night. The flow during the day is considered "free for all;" anyone can visit the stone spout and collect water in jars or jerry cans, as long as the collected water is sufficient to meet the basic requirements of the local people. Public baths or washing clothes have been banned at most of the stone spouts due to possibility of water contamination. Depending on dug wells, over ground or submersible pumps are used to pump water and store in the collection tanks. The water leakage is minimized through regular maintenance of the distribution system. Free access to water at the dug well, which is the case in the wells without a distribution system, is prohibited. The potential conflict among the water users of the SSCWSS is minimized by a fully participatory, transparent and flexible water collection and distribution system. The lack of information on water availability and the lack of transparency in the water distribution system and schedule in the KUKL's management are some of the reasons for the dissatisfaction of its customers.

In this case, the fee for access to water is either in the form of membership charges or cost of water by volume. Members get water at lower prices compared to non-members and also get access to higher volume of water during specific needs, if water is available. Furthermore, the price of water is much lower compared to the market price. The cost of 6,000 liters of water delivered by private tankers is about NRs 1,800 (US$ 24), NRs 3.33 (US$ 0.044) per liter and the quality of the water delivered by tankers is questionable. At the SSCWSS it is NRs 3 (US$ 0.04) per liter, and normally of higher quality than the water supplied by the tankers. Moreover, according to the people, the volume of water delivered by a 6000 liters tanker is usually less than that.

The schedule of water supplied by the SSCWSS, and the frequency and duration of water delivered to the member families, all depends on availability of water. For example, if enough water is available, the water is supplied to each member family on rotation basis, that is, every other day for two hours; if the availability is lesser, water is supplied only for 30 minutes every other day or even every third day. Since the members are aware of the quantity of water available for distribution, they do not generally have any qualms about the system. If the water availability is less and the member families do not get enough water for a certain period, they resort to buying water from the tankers for those days.

The members of the SSCWSS are mostly youth volunteers or selected members of the local community. The selection is usually done by gathering the community, with due consideration for representation from various locations, time-availability of the member and a known or perceived fairness of the person in dealing with different social issues. Since the members are personally known by the local residents, they feel at ease in directly talking to them about their water supply problems and seek assistance whenever needed. Conversely, the members can also call upon local residents, in times of need, for aid and contributions in kind or cash. If a particular household cannot contribute cash, they have the option of contributing the equivalent amount in kind, and vice-versa.

Conflicts do arise sometimes in the SSCWSS. Based on the available information, the nature of the conflict is mostly related to unequal distribution of water and/or the unavailability of water when needed. Some even complain of unfair supply of water to a particular member household but most of the time, the conflict is resolved by mutual agreement.

Management of Dug-well-based SSCWSS at Selected Locations

The management of the SSCWSS with dug wells as the water source at selected locations is presented below. There are many similarities and some differences among the management of different SSCWSS.

Hahkha SSCWSS

A dug well located at Hahkha, Mangalbazar, in Lalitpur is the source of the water of the Hahkha SSCWSS. The local Mother's Group and the Hahkha Youth Club are jointly managing the system. Traditionally, the well was not covered and anyone had free access to the well at any time of the day. However, after the formation of the SSCWSS, the well was covered and locked for source protection. Also, to protect the water quality of the well, the practice of washing clothes and taking bath near the well has been stopped. A submersible pump is used to extract water from the well, and the water is collected into one 10,000-liter tank and one 500-liter polythene tank supplied by the Lalitpur Municipality. When the water quantity in the well is too low, the managers of the SSCWSS request the KUKL which fills the tanks free of charge.

The membership is open to all the local residents of the area at a monthly fee of NRs 200 (US$ 2.66). The communities served by the Hahkha SSCWSS are Hahkha, Jogal, Saugal, and Kutisaugal. The current number of member families is 500. Each day, every member family gets basic water requirement (40 liters) free of charge. Additional water can be purchased at NRs 4 (US$ 0.053) per 20 liters for drinking purpose. Water for other purposes, like washing and cleaning, is supplied to 50 member families each day on rotation basis; hence, each family gets its turn to extra water on every fourth day and the fee for this extra water is NRs 70 (approximately US$ 1) for 500 liters. If extra water is needed on non-scheduled days, the fee for the water is NRs 125 (US$ 1.66) for 500 liters. The collected fee is spent on maintaining the system, which includes repairing or replacing pumps and pipes, cleaning the well, and the electricity charge for distribution. Chlorine and potash is also used to disinfect this water.

Kwelaachhi SSCWSS

The Kwelaachhi SSCWSS at Chyasal in Lalitpur, named Gajalaxmi Water Supply System, was established in 2007 with the technical assistance of Urban Environment Management Society and through kind and cash contributions of local residents, after the stone spouts in that area went dry in 2005.

Water is pumped from a recent dug well located at the base of the stone spout to 5000 liter polythene tanks. Rain water is harvested to recharge the dug well and as an additional source of water. Since the well water is found to contain higher concentration of iron the water is aerated and left overnight in the tank for the iron flocs to settle down (Tuladhar 2006). The water is pumped to three 1,000 liter bio-sand filter tanks located at higher elevation. The tanks and the filter system are cleaned when the outflow rate is less than regular flow. The filtered water is collected in 3,000 liter tanks, chlorinated, and filled in 20 liter jerry cans for sale. The water quality is regularly tested for organic and inorganic contaminants and sold only after treatment. Low-quality water from a stone spout located to the north of the Kwelaachhi SSCWSS is used for other purposes like cleaning and bathing.

The price of the treated water for the member families of the Kelaachhi SSCWSS is NRs 3 (US$ 0.04) per 20 liters of water (NRs 5 [US$ 0.067] for non-members). Depending on the availability of water, each member family is entitled to buy up to 80 liters of water per day. The monthly membership fee is NRs 300 (US$ 4) per family. Home delivery service of water is available at about NRs 10 (US$ 0.13) per 20 liter jerry can; hence the system supports local employment as well. The average daily revenue from the sale of water is about NRs 700 to 800 (US$ 9.3 to 10.6) for about 5,000 liters. The system has capacity to treat up to 6,000 liters of water per day (UN Habitat 2007). The localities served by the Gajalaxmi Water Supply System (if taken as the center) are Khapinchhen (to its south), Bhindyolaachhi (to its west), Chyasah (to its north) and Bholdhokhaa (to its east). The number of households currently served is 300, serving about 1,500 local residents. The managers of the Kwelaachhi SSCWSS are planning to expand the system by better management of the available water and the female members of the system are taking lead in incorporation of sanitation system from the generated fund. Each phase of the system is transparent and the management system is participatory; there is very little scope for conflict among the water users.

Ombahal SSCWSS

The Ombahal SSCWSS is located in core area of the down town Kathmandu Metropolitan City. Until a few years ago, water supply from the public utility was sufficient to meet the daily water demand and the traditional well was nearly abandoned; the well water was used only for religious purposes. However, with the gradual decline in water supply from the tap and simultaneous increase in water demand due to rapid population increase the local residents began to resort to the well water. A dug well, which supplied plenty of water until 2004, started to go dry due to excessive water extraction and reduced recharge rate. This situation resulted in conflict among the water users. Some local residents brought several 20 liter jars and put the jar in line which prevented other residents from getting water, because by the time all the jars of a particular family were filled, the water level in the well went too low. A local management committee was formed to distribute the water equitably and to avoid conflict.

A membership fee of NRs 100 (US$ 1.33) per month is collected from the local residents. To prevent unauthorized water extraction, the top of the well was covered and locked using steel mesh and a hand pump connected to the well was dismantled. Every morning the well is unlocked and each member family is allowed to extract 2 *gagris* (about 40 liters) of water free of charge, using their own bucket and rope. No pumps are used. When the level of water in the well is higher, the members are allowed to extract up to 80 liters of water each day. The area around the well is paved with stone slabs to prevent leakage of dirty water back into the well. The membership money is spent on maintenance and cleaning of wells. After the formation of the management committee the distribution of water has been equitable, and no conflict has been reported in the last five years.

Management of Stone-spout-based SSCWSS at Selected Locations

A snap-shot of the SSCWSS management with stone spouts as the water source at selected locations is given in the following case studies.

Aalkohiti SSCWSS

The first attempt to collect the water and distribute to the local residents living in the vicinity of the Aalkwo (also spelled as Aalkvo or Alok in some recent publications) in the Ikhaachhen area started in the early 1990s by the Patan Conservation and Development Programme (Theophile and Joshi 1992). The attempt was noble but lacked community participation. The local people for whom the system was developed did not feel ownership of the Programme and did not contribute in the operation, maintenance and preservation of the system. As a result the attempt created more problems than it solved. Eventually the system became dysfunctional. In October 2003, the local residents formed Alokhiti Conservation and Drinking Water Supply Users' Committee, installed new tanks and established a formal payment system for operation and maintenance of the water supply system (Water Aid Nepal 2006).

At the Aalkohiti SSCWSS water is collected at night into two 2,000 liter tanks located just above the spouts, an underground tank, and one 10,000 liter tank located above a four-storey building. During the day time the local residents have free access to the water; however, when the water scarcity is high the management committee prohibits free access to water. Generator facility is available to pump water. The membership fee is NRs 225 (US$ 3) per month per family. The membership fee is spent on maintaining the water pumps and cleaning the stone spout. The water is supplied to the member families on rotation basis for duration of 30 minutes to one hour depending on the water availability. The service is provided to the residents of Aalko, Ikhachhen and Naagbahal area. The number of member families is 650, including the renters.

There are seven stone spouts in the vicinity of the Aalkohiti. Community water users' groups are formed to better distribute water of these stone spouts as well. For example, the Imukhel Conservation Committee collects the water from the Wasahhiti in the overhead tanks located above the spout area, which is then distributed to the local residents at the time when the spout is dry or the discharge from the spout is too low to meet the basic drinking water requirement.

Kumbheshor SSCWSS

The Kumbheshor SSCWSS located in Lalitpur is another example of a successful SSCWSS. The water from the stone spouts is collected at night into 2000 liter polythene tanks and pumped up into overhead tanks located above a four-storey building. During the day time the local people have free access to the water at the stone spout. The collected water is distributed to the local residents (member families) as per the community fixed schedule. During the dry season the flow of water in the stone spouts is low and the piped water supply system also runs dry. Hence, the pressure on the stone spouts is high. During such times all the water discharged from the stone spout is collected in the overhead tanks and distributed to the local residents as per the schedule. The amount of water provided to the local residents on a particular day depends on the water availability.

Mangalbazar SSCWSS

The stone spouts at Mangalbazar, Lalitpur are one of the finest examples of ancient Lichchhavi period stone craft. The flow of water in these stone spouts is much higher than in the stone spouts in the vicinity. The local residents have free access to water during the morning and the day time. However, conflict started to arise especially during the water scarce season. A local community club, named Mangahhiti Sudhar Samiti, was formed to provide equitable access to water. One person from the club ensures that no water queues are broken and no family gets more than their fair share of water, especially in the water scarce season (March–May). In the morning the water is free for local residents who bring with them a gagri or a bucket, however, during water scarce season NRs 1–2 (US$ 0.013–0.026) is charged per jar. Normally, for filling a 20 liter jerry can, NRs 2 (US$ 0.026) is charged. The average daily fee collected from the water users is NRs800 (US$ 10.6) in the water scarce season. There are no water collection tanks at the Mangalbazar SSCWSS. The collected fee is spent on social works, including cleaning of stone spout area. The management style of the Mangalbazar SSCWSS is different from other SSCWSSs in the sense that no water collection tanks are used; there are no water distribution schedules and membership provisions.

The following are some of the common features at the study locations.

(*a*) At each of the locations studied, the distance between the water source and the households being served is within 1 kilometre radius. This close proximity facilitated easier access to water.

(*b*) The leadership role in most of the SSCWSS is shared by both men and women. The latter are active in many SSCWSSs. At Hahkha SSCWSS the Mother's Group is active in scheduling and apportionment of water to each household. At Kwelachhi SSCWSS, a middle aged lady is responsible for financial management of the system, three female members are in the user committee and the other local ladies play part in water distribution schedule; male members are responsible for regular checking of water quality and cleaning-up of the filtration system and water tanks. At Ombahal, Aalkohiti, and Kumbheshor SSCWSSs female members are actively involved in water distribution scheduling. Male members are expected to be responsible for heavy and mechanical works, like lifting of water pumps and hoses, repair and maintenance of electrical connections, etc.

(*c*) The price of the water is kept reasonably affordable and much lower than the market price of drinking water. The average price of 20 liters of water is about NRs 3 (US$ 0.04) which is affordable by almost all of the local residents.

(*d*) The unit of sale or access to water is kept low to maximize flexibility to both the system and to the users. Based on the requirement, a user may buy only one unit (20 liters) of water. Similarly, when water availability is low the managers of the system can provide only a few units of water to each household.

(e) Each of the SSCWSSs is free to make and change the rules based on voting during regular meetings. However, persuasion rather than voting is more in practice. Since all the members of the system are from the same locality they know each other and tend to use personal relations before formal voting to make and change the rules.

(f) Each member household was allowed access to the whole system, including water collection tanks, financial records etc. This system enhanced the trust among the members and managers of the SSCWSS.

(g) When there is too little water to be distributed among the members, each SSCWSS contacts the KUKL for assistance in filling their water tanks. Normally KUKL fills such type of community water tanks free of charge. This provision indicates the amicable relationship between the SSCWSS and the KUKL.

Lessons to be Learned

Some specific lessons to be learned from this study are summarized here.

First, when plenty of water is available for distribution (potential supply greater than demand) an agency managed water distribution system seems to work fine. Prior to the 1970s the customer base of the agency-managed water supply system in Kathmandu was small, and whenever there was no running water in the domestic pipe, people could rely on the water from the stone spouts or the dug wells. The cases of customer dissatisfaction in Kathmandu prior to 1970s were unheard of. When the supply is less than the demand conflict begins to arise in agency managed (whether private or public) water supply system due to low level of transparency in availability and distribution of water. As such, representatives of the local water users need to be incorporated in the public water utility. However, given the current nature of such types of organisations in Nepal such steps are unlikely to be taken in the near future.

Second, the experiences of the privatization of water supply system in different developing countries provide little enthusiasm. The large public utilities put low priorities in supplying water to poor sections of the settlements, regardless of whether it is an urban settlement or a rural settlement. For example, in the Kwelaachhi area in Chyasal, Lalitpur, 32 percent of the people are "too poor" and 49 percent are "poor" (Tuladhar, 2007)[3]. Private sector investment in water supply in such areas is very doubtful. Public utility in Kathmandu has not been paying attention to the plight of the people of this area for decades. As discussed earlier, a significant proportion of people's water needs here were resolved by the involvement of the local people, with the best utilization of local resources.

As such, large scale privatization of water supply system of Kathmandu Valley is not going to be feasible, at least in the current scenario. Improvement in demand management will alleviate the water scarcity problem of Kathmandu Valley to a certain extent. Improvement in efficiency of KUKL in terms of reducing the water leakage will have a very positive impact in water distribution.

Third, the direct involvement of the end users in decision making process normally results in better efficiency and sustainability of the system. The local residents are involved in formation of the SSCWSS, construction of the infrastructure, purchase of the materials and scheduling of the water distribution. As a result the dissatisfaction level, and consequent conflict, among the member households towards the SSCWSS is very low even when water availability is low.

Fourth, the direct stake of the members in maintenance of the system makes the system highly efficient. For example, in the SSCWSS the pipes for the water coming to a particular household are normally maintained by the same household. As such, there is almost no leakage in the water supply system. Water leakage in KUKL's pipes is seen at various locations in Kathmandu Valley, but few bother to report to the KUKL about the leakage since the leakage at a particular location is not directly impacting a particular household.

Fifth, as per the prevailing estimates, the existing stone spouts can serve the water needs of only about 10 percent of the population.[4] The communities in the urban and peri-urban areas of Kathmandu are coming together to construct new stone spouts and dig new dug wells and tube wells. This practice may expand the water access to more people. However, rampant extraction of groundwater can create water security problem in the long run. Luintel (2009) reported that there is a potential of acute groundwater shortage in Kathmandu

Valley in the next 100 years. Instead of extracting groundwater at ever higher rate the better option is to improve the management of both the supply and demand of the water by incorporating, among other things, the natural and artificial recharge of groundwater, extracting the groundwater at sustainable rate, and improving efficiency of various water usages, including agricultural water efficiency. This type of demand management is beyond the scope and capacity of SSCWSS. Various line agencies, including the public water utility should co-ordinate with stakeholders to manage water demand.

Sixth, the SSCWSS is not a replacement to public water utility. However, in marginalized and poor communities in urban areas of Kathmandu the SSCWSS has demonstrated remarkable success in conserving and utilizing traditional sources, generating water and managing water distribution system on their own, at affordable price, with minimum external interference. Such local initiatives need to be encouraged and assisted whenever required, in order to enhance access to water for all.

Conclusions

With the rapid rise in population, limited sources and the change from lower to higher water demanding life style, the water scarcity problem is rapidly growing in Kathmandu Valley. The government of Nepal is trying to supply more water through increase in groundwater pumping and better management of the available water. Recently a semi private company, named KUKL, is given responsibility to improve the situation; however, with no new sources of water and constantly increasing demands, the water distribution situation has not improved. The local communities are coming together to form water users' groups and revive ancient water distribution systems in which the water is tapped from distant springs and collected in water ponds. The pond elevates the local groundwater table of shallow aquifer and recharges the dug wells. Using clay pipes and brick canals the groundwater is discharged through stone spouts. These dug wells and stone spouts have been the source of water for almost 10 percent of the Kathmandu Valley residents. However, with the introduction of the modern piped water system the traditional sources of water got neglected. Consequently, the physical condition and the water quality of these traditional water sources was degraded.

With no improvement in the water supply system and no prospect of improvement in the situation any time soon, the local small-scale-community-based water supply system (SSCWSS) is taking lead in partly solving the problem of water scarcity of specific areas in Kathmandu Valley neglected by the KUKL.

Each of these SSCWSSs have revived the traditional water sources in their locality and improved the water collection and distribution system. Since the new system is participatory in nature the local residents feel ownership of the system. A nominal fee is charged to the members and other water users of the SSCWSS for the sustainability of the system. The transparency in the amount of water available for distribution and the fixed schedule of water distribution to the member families have resulted in minimizing the conflict among the water users. Although the amount of water provided to each family by these SSCWSSs is low, the members seem to be satisfied, perhaps due to equitable distribution. Each SSCWSS serves anywhere between 300 to 800 households, or families. The study results show that each of these SSCWSSs is a success story. There has been gradual increase in the formation of new SSCWSSs. Many stone spouts and public dug wells are yet to be managed in Kathmandu; hence there is a big potential to expand the system of SSCWSSs in the Kathmandu Valley.

Based on the overall performance of the SSCWSSs, it can be concluded that in the local scale within limited areas these SSCWSSs are a viable alternative to the centralized water supply system in Kathmandu. However, these SSCWSSs cannot replace the centralized water supply system due to its inherent nature of limited size and scope.

Notes

1. The Rana Dynasty ruled Nepal for 104 years (1846–1950). Movements of the people were strictly controlled during their rule. After the overthrow of the Rana Dynasty and the establishment of democracy, people from other parts of Nepal migrated to Kathmandu, mainly due to better facilities and higher opportunities. As a result, the population of the Kathmandu Valley increased from half a million in the 1950s to almost 3 million in 2011.

2. Membership is provided based on history of the house owner. Normally, any family with house ownership in the locality qualifies

for membership. Families with a long history of renting houses or rooms (usually more than five years) in the locality are also considered as local residents. The newcomers (with residency of less than five years) are not provided direct membership. However, there is no hard and fast rule in membership distribution. Even the newcomers who contribute significantly, in cash or in kind, in the initial phase of the implementation of the program, are eligible for membership.

3. The "poor" and "too poor" classification is based on household survey conducted by the Urban Environment Management Society in 2007. Families with income enough to feed (three meals a day) for three months or less in a year are considered "too poor", income enough to feed 3 to 6 months a year are considered "poor".

   The Urban Environment Management Society, Saugal, Lalitpur, Nepal, with support from UN-HABITAT, conducted an interaction program on December 17, 2007. Mrs Guheshori Tuladhar presented a paper at the program, with data on the number of "poor" and "too poor" people living in the survey area. The data is available, but not in published format.

4. The total potential of the water flow from the stone spouts in Kathmandu, Lalitpur and Bhaktapur is estimated to be 18 to 20 million cubic meters per day, which can serve the basic water needs of about 3,00,000 people. This number is approximately 10 percent of the current population of Kathmandu Valley (Spaces 2008).

References

Budds, Jessica and Gordon McGranahan. 2003. "Are the Debates on Water Privatization Missing the Point? Experiences from Africa, Asia and Latin America," *Environment and Urbanization*, 15(2): 87.

Diwakar, J., K. D. Yami and T. Prasai. 2008. "Assessment of Drinking Water of Bhaktapur Municipality Area in Pre-monsoon Season," *Scientific World*, 6(6): 94.

Haack, B. N. and A. Rafter. 2006. "Urban Growth Analysis and Modeling in the Kathmandu Valley, Nepal," *Habitat International*, 30(4): 1056–65.

Khadka, M. S. 1993. "The Groundwater Quality Situation in Alluvial Aquifers of the Kathmandu Valley, Nepal," *AGSO Journal of Australian Geology & Geophysics*, 14: 207–11.

Luintel, A. R. 2009. "In 100 Years, Valley Underground Water will Cease to Exist," *The Himalayan Times*, October 26. Available at http://www. ngoforum.net/index.php?option=com_content&task=view&id=7484 (accessed April 9, 2013).

Nepal News. 2010. "Drinking Water Being Supplied without Treatment in

Kathmandu," February 3. Available at http://www.nepalnews.com/home/ index.php/business-a-economy/3924-drinking-water-being-supplied- without-treatment-in-kathmandu.html (accessed March 5, 2013).

Gorakhapatra. 2009. "Better to Conserve Stone-spouts than Dreaming of Melamchi," June 27. Available at http://www.ngoforum.net/ index.php?option=com_content&task=view&id=6398&Itemid=6 (accessed April 7, 2013).

Non-Governmental Organizations Forum for Urban Water and Sanitation (NGOFUWS). 2006. "Traditional Stone Spouts Enumeration, Mapping & Water Quality," in *Water Movements in Patan with Reference to Traditional Stone Spouts.* Kathmandu: UN-HABITAT Water for Asian Cities Program Nepal.

Pokhrel, Damodar, 2011, "World Water Day: Water for People of Kathmandu Valley." Available at http://www.nepalnews.com/archive/2011/others/ guestcolumn/mar/guest_columns_11.php (accessed April 7, 2013).

Poudel, Keshab. 2003. "Drinking Water: From Stone Spouts to Bottles," ECS Nepal. Available at http://www.ecs.com.np/feature_detail.php?f_ id=312 (accessed March 5, 2013).

Rauniyar, Ishwar. 2010. "KUKL to Get New Tankers, Generators", *The Kathmnadu Post* (TKP), January 11.

Regional Resource Centre for Asia and the Pacific (RRCAP). 2001. *Nepal: State of the Environment Report: Water Pollution.* Asian Institute of Technology, Thailand. Available at http://www.rrcap.ait.asia/pub/soe/ nepal_water.pdf (accessed April 9, 2013).

Shakya, B. 1994. *Swonigahya Lohanhiti* (in Newari language). Lok Sahitya Parishad. Nepal

Shrestha, Roshan. 2009. "Rainwater Harvesting and Groundwater Recharge for Water Storage in the Kathmandu Valley," *Sustainable Mountain Development,* 56: 27–30.

Spaces. 2008, "Hitis: An Alternative Source of Water," http://www. spacesnepal.com/archives/mar_apr08/hitis.htm (accessed April 9, 2013).

Theophile, E. and P. R. Joshi. 1992. *Historical Hiti and Pokhari: Traditional Solutions to Water Scarcity in Patan.* Unpublished report. Patan Conservation and Development Programme, Urban Development through Local Efforts, GTZ, Gutschow Hagmuller and Associates, Germany.

Tuladhar, Guheshori. 2006. "Community-based Clean Water Bottling System" (in Nepali language), Urban Environment Management Society, Lalitpur, Nepal

UN-HABITAT. 2008. *Mapping the Footprints of Water Movements in Patan.* Kathmandu: Water for Asian Cities Program.

———. 2008. *Safe Drinking Water for Urban Thirst.* Community-based Safe and Treated Drinking Water Bottling Plant in Chyasal. Kathmandu: Water for Asian Cities Program.

Warner, N. R., J. Levy, K. Harpp, and F. Farruggia. 2008. "Drinking Water Quality in Nepal's Kathmandu Valley: A Survey and Assessment of Selected Controlling Site Characteristics," *Hydrogeology Journal*, 16: 321–34.

Water Aid Nepal. n.d. Available at http://dev.www.wateraid.org/international/what_we_do/where_we_work/nepal/5734.asp (accessed April 7, 2013).

Welle, K. 2006. *Water and Sanitation Mapping in Nepal*. A Water Aid Report.

9

Need for Reforming the Reform

Incompatibility and Usurpation of Water Sector Reforms in the Indian State of Maharashtra

Sachin Warghade and *Subodh Wagle*

—

The reforms in the water sector need to be seen in the context of the overall discourse of "good governance," which came in to prominence in the context of "Globalization" (Chakrabarty and Bhattacharya 2008). Along with the conditions on the administration of the aid, International Financial Institutions (IFIs), such as the World Bank (WB), while providing financial support for development projects also included the conditions for bringing reforms in the governance framework (Mathur 2005, 2008). According to the WB (1992: 1), "good governance is central to creating and sustaining an environment which fosters strong and equitable development, and it is an essential complement to sound economic policies." Excessive regulations which impede the functioning of markets and the absence of a predictable framework of law and government behavior are listed by the WB as some of the important symptoms of poor governance (ibid.).

Critiques, however, point out that curbing the role of the State and expanding the space for market and competition is the main agenda behind good governance, leading to advocating ideas such as downsizing the government, privatization, contracting services, decentralization and other aspects, proposed as part of reforms (Mathur 2008). One of the major pitfalls is the "tendency to 'depoliticize' government and bring in more technicism and expertise at the cost of citizens' age-old and hard fought democratic right to govern politically" (Chakrabarty and Bhattacharya 2008).

In the light of this critique, the IFIs revised the discourse from "minimal" State to "effective" State.[1] The role of the government in matters of governance, such as policy-making, creating rule of the law, and developing strong institutions, was seen as crucial for economic and social development. The WB's view on "governance" also began to emphasize upon principles of "equity" and "inclusiveness" and includes such principles that look different from its conventional "technical" outlook (Bhattacharya 2008). So, the question that arises is whether the reforms, undertaken under the influence of the IFIs, are actually compatible with their principles of equity or inclusiveness. This chapter attempts to address this question based on an assessment of some of the latest reforms in the water sector.

Factors for the Assessment of Reforms

The sectoral reform measures, as promulgated by the IFIs, are justified on account of the various challenges and problems plaguing the water sector; for example, water sector reforms are expected to provide "specific, practical solutions to local problems" (World Bank 2005: 65). Reforms are seen as a response to the serious breakdown in services or a fiscal crisis which makes the status quo untenable (ibid.: 63). They are also regarded as a response to the expectation of the stakeholders, especially the water users. For example, the rules for initiating reform include–"Initiate reform where there is a powerful need and demonstrated demand for change. Involve those affected, and address their concerns" (ibid.: xiv).

The progenitor of sectoral reforms, the World Bank, has created a significant number of documents prescribing different norms and standards for the content as well as the process of sectoral reform. One such document prescribes the following principles for all infrastructure regulatory systems: Credibility, Legitimacy, and Transparency (Brown et al. 2006: 7). It further elaborates this theme in terms of "a number of institutional and legal principles" including "accountability, transparency and public participation, predictability, clarity of roles, completeness and clarity in rules, proportionality, requisite powers, appropriated institutional characteristics, and integrity" (ibid.).

Thus, the measures of sectoral reform in the water sector need to be seen as a response to two factors: (*a*) needs that emerge from the current problems and challenges faced by the sector, and (*b*) expectations of the water users, who are the worst victims of the problems plaguing the sector. Further, the reform measures in Indian states rely heavily on the financial assistance provided by public-funded agencies such as the Asian Development Bank and the World Bank. As a result, they are also expected to adhere to the norms and standards of these organizations.

Therefore, the scheme of analysis devised for this chapter (see Figure 9.1) is to try and assess the sectoral reform measures by evaluating whether they fulfill the demands created by (*a*) sectoral needs, (*b*) the expectations of water users, and (*c*) the norms of the World Bank.

Contextualizing the Assessment

While contextualizing the research in actual reform measures, the first step was to choose and focus on a particular state, as sectoral reforms in water is a state-specific phenomenon. The state of Maharashtra was the obvious choice, as it is the pioneering state in India in terms of sectoral reforms. The next step was the choice of particular reform instruments. Maharashtra is also the first state to enact the law for the establishment of an Independent Regulatory Agency (or IRA) in the water sector in 2005. The Maharashtra Water Resources Regulatory Authority (MWRRA) Act, 2005, defines the design of the structure and functioning of the IRA. It is one of the most comprehensive and powerful reform instruments prescribed by the World Bank and other mainstream agencies. Similarly, the state has made progress in the area of privatization of irrigation projects through corporate investments and management and introduced a policy instrument in the form of a Government Resolution (GR), in 2003, which stipulates the policy of privatization of incomplete irrigation projects (PRAYAS 2003); for example, the Nira Deoghar Irrigation Project was one of the first projects of its kind.

Thus, the assessment in this chapter is contextualized in two sets of reform instruments or measures (see Figure 9.1). The first set comprises those which define the policy and procedures for the privatization of incomplete irrigation projects while the second set comprises those which pertain to the establishment and functioning of the IRA. For the purpose of their assessment, the design of the individual reform instruments as well as their implementation has also been considered. The various regulatory interventions initiated by the authors of this chapter have contributed in gaining understanding about the execution of these instruments.

Findings and Analysis

The findings related to compatibility are presented in this section for the following reform measures selected for the assessment–privatization of the irrigation projects and the establishment of IRAs.

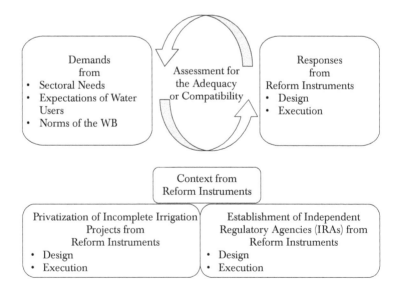

Figure 9.1: Scheme of Analysis

Source: Prepared by the authors.

Compatibility in Reform Measure 1: Privatization of Incomplete Irrigation Projects

The water sector in Maharashtra required a credible, implementable and viable plan to raise the huge funds, about INR 36000 crore (US$ 9 billion), for the completion of the project[2]–as a result of the combination of two compelling factors: (*a*) the ever-widening gap in the demand and supply of water, and (*b*) the large number of incomplete irrigation projects (about 1200) requiring infusion of capital. The government attempted to respond to this need by establishing a sectoral reform measure following the route of privatization on the "Build, Operate, and Transfer" (BOT) basis.[3]

Compatibility with Respect to the Demand for Raising Finances

After introducing the GR (2003), a "High Power Committee" (HPC) was constituted in 2006 under the direct chairmanship of the Chief Secretary, Government of Maharashtra for spearheading the process of privatization and evolving plans for attracting private investors (PRAYAS 2009b). The Nira–Deoghar (ND) project was chosen as the pilot project to be completed through the privatization model and this initiative was officially approved by the HPC in 2007. An advertisement for "Expression of Interest" (EOI) was published by the Maharashtra Krishna Valley Development Corporation (MKVDC), a government-run corporation. Following the advertisement, a conference and study visit of prospective investors was organized.

The proposal prepared by the MKVDC was based on the privatization policy elaborated in the GR (2003). The comparative analysis of the provisions in the GR and the relevant provisions in the MWRRA Act 2005 showed many contradictions and discrepancies. For example, the privatization policy as per the GR allowed the private investor to recover his/her capital investment through the revenue generated from water charges levied on water users as well as from other sources, such as contract farming, fishing rights, and tourism. Similarly, the GR allowed the private investor to decide the allocation of water during the scarcity period. However, the MWRRA law does not permit recovery of capital costs from water charges and has the authority to decide upon water allocation in the scarcity period, unlike the GR.

The regulatory interventions based on this contradiction resulted in a judicial order by the MWRRA demanding withdrawal of the proposal for privatization.[4] The order held that the provisions in the GR (2003) were null and void in view of the provisions in the MWRRA Act, 2005. Accordingly, the advertisement for the EOI was withdrawn by MKVDC and the process was stalled until the time the government was able to revise the policy.[5] Thus, reform measures based on the principle of privatization were found not only internally contradictory, but were also inadequate to fulfil the serious and urgent need of the water sector, that is, evolving a viable, legally sound and politically acceptable plan for financing the completion of incomplete irrigation projects.

Protection of Rights of Water-Users

The privatization model was expected to be planned in a way that did not compromise the rights and benefits of water users, including the local farmers and other beneficiaries. During official deliberations over the Nira–Deoghar Build, Operate and Transfer (ND–BOT) project, it was found that revenue generation from sources other than water charges would not be substantial and sustainable. In the light of the clear legal ban on the recovery of capital from water charges, the only way left for generating substantial revenue was to divert the water from agriculture to more revenue-generating sources, for industrial and other commercial purposes (PRAYAS 2009a). This posed a serious threat to the interests of the farmers and there seemed to be no protection to them against this possibility in the reform instruments.

Furthermore, the GR (2003) gave the private developer the authority to control the operation and management of the irrigation system on the dam–in contradiction with the provisions of the Maharashtra Management of Irrigation Systems by Farmers (MMISF) Act, 2005, another reform instrument applicable to the Nira–Deoghar project. This was seen as another threat to the benefits and rights of farmers because, as per the MMISF law, it is the right of the farmers to take over the complete management and operations of the irrigation system. Therefore, it seems that the reforms related to privatization were designed and implemented in a manner that was incompatible with the expectations of the beneficiaries.

Adherence to Principles of Participation and Accountability

The World Bank (2009) prescribes detailed and strict norms and standards for adhering to the principles of meaningful and true participation and effective accountability while designing reform instruments and their implementation. However, there had been no effort for confidence-building and catering to public concerns by any government agency, involved at any stage of preparing the GR (2003), for privatization.

While implementing the GR, in the case of ND-BOT, there was no public consultation; neither was there any information available on the websites of the concerned agencies nor was any data available in the public domain about the booklet created for the EOI procedures. There was no plan to invite objections or suggestions from the prospective beneficiary farmers of the dams or other stakeholders. In fact, the government agencies, during the hearings on the petition, fought intensely against the suggestion of public participation in the decisions regarding the privatization of the ND-BOT (PRAYAS 2009a).

Usurpation and Bypassing of Powers of MWRRA

It was found that neither the HPC nor the MKVDC made any attempt to involve or consult the MWRRA on the issue of privatization. In a response to the petition filed against the privatization process, the MKVDC (in a very strident and bitter manner) tried to browbeat the objections, arguing that the HPC was an appropriate and competent body to take all decisions and that the intervention of the regulator was not required. It is clear from the statement in the written response from the MKVDC that it was a conscious decision not to bring the privatization issue under regulatory purview, instead of a lack of awareness or ignorance. In other words, it was a systematic effort to usurp the powers and authority of the MWRRA to avoid an independent and transparent scrutiny of the privatization proposal (ibid.).

While all these developments were taking shape, there was no immediate response from the farmers or other rural water users (whose rights were under threat) to the advertisement for privatization, though they had raised concerns about this during the

WB visit conducted much before the advertisement was released. They also failed to make any intervention at the time of the conference with prospective private developers or their field visits, arranged by government officials. Since they were not made aware of this privatization process, the farmers remained alienated (ibid.).

Compatibility in Reform Measure 2: Establishment and Operationalization of IRA

The analysis of the MWRRA law and its implementation is assessed in this sub-section on the basis of various key expectations and demands.

Contravention of Principles of Transparency, Accountability and Participation

The norms and standards prescribed by the World Bank require that the regulatory reform should bring in the key principles of "Good Governance"–transparency, accountability and participation (TAP)–in the functioning of not only the state agencies but also of the new regulatory agencies. Hence, it was expected that the law that prescribes the design of the structure and functioning of the new water sector IRA (namely, the MWRRA) should strive to ensure that the law would set up new, higher levels of standards by improving upon the standards of the TAP, prescribed by previous laws, such as the Electricity Act, 2003. However, a detailed comparison of the two laws–the Electricity Acts 2003 and the MWRRA Act, 2005–revealed that the latter was very weak and deficient (in TAP standards) as far as its structure and functioning was concerned (PRAYAS 2009b).

When it comes to the implementation of the MWRRA law, there are instances of equally serious contravention of the very principles of "Good Governance," mentioned earlier. First, the agency has not prepared the Conduct of Business Regulations (CBRs), even after four-and-a-half years of its establishment. It needs to be noted that, as in the case of the other acts, the government does not prepare the rules that provide detailed guidelines for the functioning of the IRAs. In fact, the IRAs are allowed to prepare the regulations that govern their own functioning and these are known as the CBRs. In their absence, the only framework for the functioning of the IRA is the law which creates it and this simply provides a broad framework. In the absence of all rules or regulations, the IRAs have no benchmarks to follow while conducting their own business, thus, becoming the law unto themselves (ibid.).

This is what has happened with the MWRRA; the authority has consistently refused to frame the CBRs, despite repeated reminders and requests. This allows the MWRRA to interpret the law as per its own convenience and change even the decisions about its functioning (self-made), itself. For example, the MWRRA flouted and has allowed private consultants to flout the Terms of Reference (ToR) it set itself, changing the pre-decided process of preparing and finalizing a critical document like the Tariff Approach Paper (PRAYAS 2009b). In fact, the MWRRA has not prepared any regulation that is required by the law, leading to contravention of the law itself. For example, even though the law clearly mentioned that the MWRRA should appoint consultants as per the regulations to be determined by the regulator itself, it has appointed consultants for many tasks without even preparing the regulations governing their employment. Moreover, the MWRRA has been conducting reviews and clearance of projects, without having devised any policies for this process.

The composition of the MWRRA is another example of contravention. The law prescribes that the authority should have, as its members, an expert each from fields of engineering and economics, who are generally bound by their accountability to peers, while making technical decisions. However, these positions of authority have been typically occupied by retired government officials. This is easily facilitated as the membership of the purportedly independent "Selection Committee," that chooses the members of authority, is completely monopolized by the serving government officials. In fact, a member (with an engineering background) appointed on the MWRRA in 2009, after the end of the term of the first member, was a person who had just retired from the post of the Secretary of the Water Resources Department (WRD). It is ironical that this same Secretary of the WRD also was the Member Secretary of the Selection Committee, since the law stipulates that the Secretary of WRD shall be the *ex-officio* Member Secretary of the Selection Committee. Thus, the person who initiated the proceedings of selecting members for the post of the Secretary (by the Selection Committee) was himself selected for the same post (GoM 2009a, 2009b). This not only destroys the credibility of the institutions, but also undermines the principle of peer-based accountability.

Moreover, such appointments also end up creating an undesired nexus between the government departments (to be regulated) and the regulator.

Refusal to Regulate Key Financial Aspects

The stakeholders (especially the water users) had expected that the IRA would tighten the control on financial matters and reduce corruption and other irregularities.

Tariff is a crucial tool to regulate financial matters of utility, as it gives a handle to the regulator to control investments and capital expenditure, which have serious implications for the tariff levied on consumers. However, the law does not provide for explicit powers and jurisdiction to the regulator over capital expenditure which is a major flaw, as far as this particular demand is concerned. This has been brought to the notice of the MWRRA, but it fails to envisage any role in ensuring efficiencies in capital investments and expenses, citing a very narrow interpretation of the law. It argues that as the law restricts calculations of water tariff only to the extent of the recovery of operation and maintenance (O&M) costs and not the capital cost, it has no jurisdiction over capital expenditure. It is not convinced by the argument that capital investments do have implications for tariff via several routes–especially that the quantity and quality (and/or efficiency) of capital expenditure would affect all O&M expenses. The latter would also affect the total quantity of water that can be made available within the available capital resources.

Thus, in effect, the reform law and its implementation fail to fulfill the most basic requirement of regulating affairs of any utility, which is regulating their capital expenditure.

Failure to Ensure Techno-economic Accuracy and Rationality

The core expectation from the IRA was that it would ensure technical accuracy and techno-economic rationality in its decisions, as the basic motivation for creating such an IRA is the premise of "rational decision-making" in the sector.

Unfortunately, there have been numerous conceptual, methodological and data-level lacunae in the Tariff Approach Paper and its related processes. The Approach Paper prepared by the consultant for the MWRRA, on a critical issue like water tariff, was replete with numerous technical errors (PRAYAS 2009b). The MWRRA failed to undertake a critical review of this document, thereby landing itself into embarrassing situations during the public consultations on this Approach Paper.

Planning and implementation of water projects is one of the important areas where techno-economic accuracy and rationality of decision would play a critical role. The MWRRA law requires the Integrated State Water Plan (ISWP) to be prepared and finalized within six months of passing of the reform law, according to which the MWRRA has to clear new projects based on the finalized ISWP. Nonetheless, the MWRRA has been reviewing and clearing projects (in absence of the ISWP), without devising any explicit criteria for review/clearance, and without any actual field investigations (MWRRA 2010). Therefore, it has become an instrument in continuing and legitimizing irrational processes and decisions, even after the passing of the reform law. On the whole, the MWRRA has failed to ensure that its functioning or decisions abide by technical accuracy and techno-economic rationality.

Failure to Protect Interests of Poor and Vulnerable

Regulating water tariff and distribution of water in the form of water entitlements are two crucial regulatory functions entrusted by the law with the MWRRA. It is assumed that the reform measure would ensure protection of the interest of the poor and vulnerable sections in both these crucial functions.

In the case of water tariff, there is no explicit provision in the law for ensuring "affordability" and "equity." This led to the assumption by the MWRRA-appointed consultant that the tariff should be such that it reduces "cross-subsidy." Affordability and equity principles were completely ignored in the Approach Paper prepared for the MWRRA (PRAYAS 2009b). It was only after strong, and adversarial, interventions by civil society organizations that the MWRRA had to change the Approach to include various concessions, based on social considerations (PRAYAS 2011). For entitlements and distribution of water, there is no provision for water entitlements to the landless (and other vulnerable sections) in the MWRRA law. Water entitlement continues to be given on the basis of land-holding—those who hold bigger lands get more water while landless people get none. This regressive distribution of the crucial resource, created through public funding, is a serious failure of the reform measure in protecting the interests of the vulnerable classes (PRAYAS 2009c).

Further, the MWRRA law allows trading of entitlements. First of all, the inequitable distribution of initial entitlements permanently rules out the poor and the marginalized from the markets of water entitlement. It is feared that, in absence of other measures to protect small farmers, the market would gradually lead to the curbing of water by the bigger corporate players, to the permanent exclusion of the poor and vulnerable sections of society. A study based on the data collected through the Right to Information (RTI) Act revealed that, since 2003, a ministerial group, led by the then Water Resources Minister, diverted huge amounts of water from agriculture to non-agriculture use. More than 1,500 million cubic meters of water, from 38 irrigation projects, was diverted for industries and urban cities (PRAYAS 2010). This was done after the establishment of the MWRRA, when it had the powers to decide water entitlements. The ministerial group bypassed the regulatory provisions related to water entitlements.[6]

This is one of the critical failures of the MWRRA to protect the interests of the poor and marginalized sections of society, even though it had been given the responsibility to determine the entitlements in an "equitable" and transparent manner, and highlights the attempt to usurp and bypass the regulatory reforms for ensuring water to the urban-industrial sector, at the cost of agriculture.

Refusal to Monitor Performance, Control Costs, and Ensure Efficiency

According to its basic function as an IRA, the MWRRA is expected to take up the tasks of monitoring different aspects of performance, reducing losses, theft and wastages, controlling costs, and increasing the efficiency of performance. The preamble of the law talks about ensuring sustainable, judicious and equitable management of water resources as the main objective. But there are no concrete provisions for guiding and requiring actions to this effect. In fact, the MWRRA is determining and regulating tariff, even when there are no adequate systems in place for measuring water use and water losses (PRAYAS 2009b).

Further, though it accepts the responsibility for ensuring the recovery of the O&M costs through tariff, the MWRRA does not envisage any role in ensuring that the expenditure is made in a productive and competent manner. It has also consistently refused to take responsibility for ensuring the efficiency of the O&M and/or monitor its outcome. Thus, the MWRRA is trying to implement crucial principles, such as "full cost recovery" (of O&M costs) and "tradable entitlements," without actually developing proper and adequate systems for the measurement of data, monitoring and control of costs and reduction of losses.

Conclusion

The reform measures were found to be incompatible and inadequate on different counts with the demands of the water sector, stakeholders and the norms prescribed by the World Bank (see Figure 9.2). In cases like the privatization of irrigation projects, the reform instruments were not compatible with each other. Even the provisions, given in the different instruments, contradicted each other. Furthermore, the responses of different stakeholders to all three sets of incompatibilities and inadequacies demonstrated two dominant trends: (*a*) on one hand, the dominant stakeholders–including the government, the private industry and the consultant–seemed to be making concerted efforts to bypass the MWRRA and the provisions of the MWRRA law as well as encroach upon the rights and benefits of the weak stakeholders, to secure and garner the benefits for themselves or for their clients; (*b*) on the other hand, the weak stakeholders appear to be removed and alienated from the regulatory processes and fail to make any effort or interventions to secure their legitimate rights and benefits (see Figure 9.2).

This second trend could be explained by the near-complete ignorance and lack of awareness (about the new law and establishment of the new IRA) on the part of the larger section of the citizenry, and most sections of the urban society. They are completely unaware of the different decision-making processes (such as the privatization of projects or determination of water entitlement and/or tariff) as well as about the regulatory mechanisms that are expected to protect their rights during these processes.

In fact, these two trends could be linked and, consequently, it could be argued that the alienation of the weak stakeholders has allowed the dominant groups to try to usurp maximum benefits, as discussed earlier. Moreover, the alienation of the weak stakeholders and their absence in the regulatory processes could be traced to

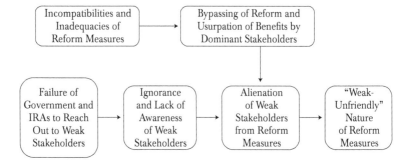

Figure 9.2: Different factors connected to incompatibilities and usurpation

Source: Prepared by the authors.

the failure of the government agencies and the MWRRA to reach out to these consumers and inform them about the possibilities and opportunities provided by the new reform instruments. At the same time, some researchers also try to explain the causes of such alienation by arguing that the new regulatory regime, created by the sectoral reform, is distinctly unfriendly to the weak stakeholders. This characteristic is said to be rooted in different elements of the new regulatory regime, especially the "expertocratic" and bureaucratic organizational structures and the technicised nature of its functioning (Warghade and Wagle 2009).

Therefore, the assessment of the the current water sector reforms, in this context, clearly indicates their incompatibility with progressive social principles of "equity" and "inclusiveness," among others. Although the IFIs (like the World Bank) include such principles in their "revised" governance discourse, the reality is quite different from the principles being promulgated. Reform proposals for restructuring the role of the State, the regulator and the market are found to be inconsistent with progressive social principles. The 'depoliticizing' agenda has been found leading to various distortions that enable the dominant groups to continue their stranglehold on water governance, thus pointing to the need for reforming the currently proposed "reforms"–by taking due cognizance of the political nature and structures of the resource. The creation and implementation of a strong policy and programmatic framework for ensuring social objectives, such as equity, inclusiveness and democratic governance, should be the core components of such reforms. Hence, governance should be seen as "a project of continuous struggle for social construction, which includes issues of inclusion, equity, and equality" (Chakrabarty and Bhattacharya 2008).

Notes

1. Welfare state, effective state and minimal state can be seen as three levels on the continuum of state control of the economy and public affairs. Minimal state refers to the minimum intervention of the state in the private and public affairs of the citizens. It emphasizes on the importance of free market to be created through deregulation and privatization in every possible sphere of life. Effective state refers to an optimum level of intervention of the state especially with relation to its regulatory role. It also emphasizes the importance of free market but differs from the minimal state in the intensity of intervention. It assumes that the state has an important role of promotion and protection of market as well as of public interest. In contrast to these, the welfare state falls at the extreme end of state control where the state takes on the responsibility of not only regulation but also production and provision of all important forms of public services and goods.
2. The currency conversion rate of the time was INR 45 to 1 US$. The projects comprise construction of dams and/or canals.
3. BOT is a form of public–private partnership (PPP). In this case, it involved the completion of irrigation projects, especially the canal systems through investments by a private company. In such projects, the BOT period is long (30–40 years) and, almost one generation of farmers has to use the services of the private company. Bringing in private corporate investment also leads to full-scale commercialized operations. However, technically, these projects are partially completed using public funds and, hence, cannot be completely privatized.
4. The intervention was done by the authors in the form of a petition filed before the MWRRA, which is a quasi-judicial body. The detailed order, given by the MWRRA, on the petition is available on www. mwrra.org.
5. Until the time of finalizing this chapter (that is, around May 27, 2011), the government has still not been able to introduce a revised policy for privatization.
6. While this chapter was being finalized, the government passed a bill in 2011 for the amendment of the MWRRA law, in response to the legal petition filed by various groups (including the authors) against water diversion. The amendment says that all past water allocation decisions

shall be legally valid and they cannot be challenged in any court. In future, the decisions on inter-sectoral allocations will be taken by the government and not the regulator. While doing so, the government will not be bound by the principle of "equitable" allocations that existed in the original law (for details, see PRAYAS 2011).

References

Bhattacharya, M. 2008. "Contextualizing Governance and Development," in B. Chakrabarty and M. Bhattacharya (eds), *The Governance Discourse: A Reader*, pp. 79–102. New Delhi; New York: Oxford University Press.

Brown, A. C., J. Stern, B. Tenenbaum, and D. Gencer. 2006. *Handbook for Evaluating Infrastructure Regulatory Systems.* Washington, DC: World Bank.

Chakrabarty B. and M. Bhattacharya. 2008. "Introduction," in B. Chakrabarty and M. Bhattacharya (eds), *The Governance Discourse: A Reader*, pp. 1–78. New Delhi; New York: Oxford University Press.

Government of Maharashtra (GoM). 2009a. Notification for Appointment Of Members Of MWRRA under MWRRA Act XVIII of 2005. Mumbai: Water Resources Department.

———. 2009b. Appointment Notification dated 20 August 2009 (*Adhisuchana*). Mumbai: Water Resources Department.

Locussol, Alain R. and Matar Fall. 2009. "Guiding Principles for Successful Reforms of Urban Water Supply and Sanitation Sectors." Water Working Notes No. 19. Washington DC: World Bank.

Mathur, K. 2005. "Administrative Reform in India: Policy Prescriptions and Outcomes," in B. Chakrabarty and M. Bhattacharya (eds), *Administrative Change and Innovation: A Reader*, pp. 278–94. New Delhi; New York: Oxford University Press.

———. 2008. "Administrative Reforms and the Demand of Good Governance," in B. Chakrabarty and M. Bhattacharya (eds), *The Governance Discourse: A Reader*, pp. 211–30. New Delhi: Oxford University Press.

PRAYAS. 2009a. *Privatization of Incomplete Irrigation Project: Judgment of Maharashtra Water Resources Regulatory Authority on Privatisation of Nira Deoghar Irrigation Project.* Pune: PRAYAS.

———. 2009b. *Independent Water Regulatory Authorities in India: Analysis and Intervention.* Compendium of Analytical Work by PRAYAS (2006–09). Pune: PRAYAS.

———. 2009c. "Water Sector IRAs and Institutional Reforms in India," in *Proceedings of the National Workshop on Independent Regulatory Authorities (IRA) and Related Institutional Reforms in the Water Sector in India held in Mumbai*, August 28, 2009. Pune: PRAYAS.

————. 2010. *Study on Diversion of Water from Agriculture to Non-agriculture in Maharashtra.* Pune: PRAYAS.

————. 2011. "Background Notes for National Consultation on Water Regulatory Authorities in India: Rethinking the Current Models." Paper presented at the workshop held at the Indian Institute of Technology, (IIT) Bombay, April 30, 2011, PRAYAS, Pune, India.

Warghade, S. and S. Wagle. 2009. "Assessment of Reform Policy Instruments for their Contribution to Empowerment and Equity in the Water Sector," in *Proceedings of the Fourth South Asia Water Research Conference on Interfacing Poverty, Livelihood and Climate Change in Water Resources Development: Lessons in South Asia,* May 4, 2009, Kathmandu. Hyderabad: South Asia Consortium for Interdisciplinary Water Resources Studies (SaciWATERs).

World Bank. 1992. *Governance and Development.* Washington, DC: World Bank.

World Bank. 2005. *India's Water Economy: Bracing for a Turbulent Future.* Report 34750-IN. Washington, DC: World Bank.

10

Understanding Emerging Independent Regulatory Frameworks

Lessons for Reforming Karnataka's Water Governance

Divya Badami Rao and *Shrinivas Badiger*

"Water governance" refers to the entire set of systems that control decision-making with regard to water resource development and management. It covers both the manner in which decisions about the allocation and regulation of water are made, and the formal and informal institutions by which authority is exercised. How societies choose to govern their water resources has profound impacts on people's livelihood opportunities and sustainable management of water resources. Apart from being unevenly distributed in time and space, water is also unevenly distributed among the various socio-economic strata of society, in both rural and urban settlements. Livelihood opportunities of low income groups, and minorities in particular, depend directly upon sustained access to natural resources, especially since they tend to live in areas that are prone to pollution, droughts and floods. Improving water governance will, thus, provide one cornerstone to alleviate poverty.

The provision of water for irrigation and domestic use, and industry was until recently, the responsibility of the state government. However, recent water sector reforms, initiated by International Financial Institutions (Ifis), paved the way for privatization of water management and delivery services through provisions made under the Independent Regulatory Authorities (IRAs)–which have been introduced to essentially regulate the water sector in India. The state of Karnataka, though steeped in water sector reforms, does not have a "regulatory institution" as yet, but that may well be on the cards. In the first two sections, this chapter contextualizes Karnataka's water sector reforms and the birth of the IRAs in India. It will then examine some conceptual and empirical approaches to addressing issues of water governance, with respect to transparency, public participation and accountability, amidst concerns of equity and sustainability. While briefly exploring alternative governance mechanisms, it outlines key challenges of the water governance system. In conclusion, the chapter anchors the role of the government as a custodian of water resources, and the need to circumscribe water governance in the context of the "Right to Water," premised upon principles which are socially and environmentally sound, with greater emphasis on public control over scarce resources, such as water in its governance.

Water Sector Reforms and IRAs

Diagnosis of IFI's influence on framing India's water policy suggests that problems with subsidization, low cost recovery and general inefficiency in the system lead to its overhaul by attempting to make it a profitable sector (Wagle and Dixit 2006). Reforms are, therefore, predicated on market-based principles, which are accepted as part of the agenda for economic liberalization and globalization. The intentions expressed, for instance, in the World Bank Group's FY09–12 Country Strategy (CAS) for India are a cross-section of organizational and institutional changes, focussed on allocations driven by tariff systems that highlight the tone of financial viability. It is important to note that organizational and institutional changes can be introduced into the system, independent of each other (Ramachandrudu et al. n.d.). Organizational changes alter roles, responsibilities, authority, and powers of different key organizations and actors and their inter-relationships in the water sector. Institutional changes, on the other hand, are changes in basic principles, norms, rules, and procedures related to different aspects of water sector governance, such as water rights, water tariff and planning. In India, changes in governance functions were initiated even before organizational changes, such as the creation of IRAs, were introduced.

The obvious question, then, is–what role will the IRAs play, given that reform processes in most states have attempted addressing issues of financial and performance crisis even before these new sector reforms were instigated. Though "regulation" as "the administrative technology of controlling business through law-backed specialized agencies" in some form has always been a key part of the state role, IRAs are distinguished by being separated from the executive branch of government so that they function independently (Dubash 2008: 43). While subsidization of water and low cost-recovery have been singled out as the "problem" of the water sector, under political pressure from the poorer section of voters, the government has been reluctant to increase tariffs. In the absence of political constraints, therefore, IRAs can independently set tariffs, among other pertinent decisions. The government can then claim no responsibility for and remains unaccountable to the actions of the IRAs and its outcomes. As one may argue, creation of IRAs seems like a clever process of depoliticization of the water sector, to institutionalize new principles of water governance, such as "full cost recovery" or to prioritize "economic uses of water."

While the key IFIs, in the water sector in Karnataka, are the Asian Development Bank (ADB), World Bank (WB) and Japan Bank for International Co-operation (JBIC), the WB has the biggest presence, and has had the greatest influence. The WB has financed seven sector reform projects in Karnataka, while the ADB has financed three, and the fourth one (though not approved) is waiting to adhere to IFI expectations on cost recovery. The WB's argument that governments in developing countries are too poor and too indebted to subsidize water and sanitation services, are reflected in its structural adjustment and water and sanitation loans. These loans routinely include conditions requiring increased cost recovery, full cost recovery or economic pricing for water services (Grusky 2001). These changes have been enabled by altering supporting policy and legislations to adhere to the conditions upon which WB and other IFIs agree to extend loans toward water sector improvement projects. Thus, the National Water Policy (NWP), 1987, was replaced by the National Water Policy, 2002, to accommodate new elements. Drawing from this, the Karnataka State Water Policy (KSWP), 2002, includes tariff restructuring, and encourages private player participation, reviews of existing policies and formulation of new ones. The KSWP 2002 is reflected in the slew of water-related amendments and legislations issued which include the Karnataka Urban Drinking Water and Sanitation Policy, 2002; Karnataka Municipal Corporations (Water Supply) Rules, 2004; and the Karnataka Ground Water (Regulation and Control of Development and Management) Bill, 2006. The Karnataka Irrigation Act, 1965, was amended in 2000 and 2002 to encompass provisions for participatory irrigation management. While Karnataka has been quicker to introduce organizational changes through individual reform elements, institutional restructuring in the state is still in its infancy and largely controlled by the irrigation bureaucracy.

Regulatory Authorities in Karnataka

Karnataka Urban Water Supply and Sanitation Council (KUWSSC) is a proposal mooted by the Karnataka Urban Infrastructure Development and Finance Corporation, in 2004, through the Karnataka Urban Water Sector Improvement Project (KUWASIP), partly funded by the World Bank. This proposal is based on a study (KUIDFC 2007), jointly conducted by The Energy and Resources Institute (TERI), Credit Rating and Information Services of India Ltd (CRISIL) and Tahal Consulting Engineers, and is expected to be set up in the next 12–18 months. Though currently envisaged to be an advisory body, it may also be structured in the form of an "empowered council" to bring in different aspects of regulation in a phased manner. The KUWSSC, whose jurisdiction will be over 219 Urban Local Bodies (ULBs) of Karnataka, will be engaging with issues of tariff frameworks and guidelines, macro-level water supply plans, reform-based guidelines for the ULBs, ULB trainings, encouraging the private sector by creating model contracts, and assisting the government in conflicts that may arise between the Karnataka Urban Water Supply and Drainage Board (KUWSDB)– an implementing body for water supply and underground drainage schemes in urban areas of the state, except Bangalore city–and the ULBs. The composition of the board may include a chairperson, technical member, a finance member, and a public relations member, amongst others. Civil society representation on the board is not confirmed.

The second attempt at introducing an IRA for Karnataka is the Karnataka Water Resources Authority (KWRA), 2008– also envisaged as an advisory body instituted via a government order dated September 9, 2008. The KWRA, named the steward of state's water resources, has been established according to the order because "water resources in Karnataka need to be planned, developed and managed effectively, efficiently, and in an integrated manner to avoid wastage and to promote and facilitate equitable use of water" (KWRA 2008: 1). Surprisingly, the tenets of efficiency and equity were missing in the earlier missions of the Water Resources Development Organization (WRDO). However, the KWRA has precisely two concrete functions of consequence to decision-making, that is, (*a*) "to determine the quantum of water, surface and underground that can be converted to public use or controlled for public protection;" and (*b*) "to propose a water tariff system and fixation of tariff for full cost recovery in a specified time bound manner for different water uses and areas" (KWRA 2008: 3). Other functions relate to carrying out studies of water requirement, surface and groundwater hydrology, and water audits for irrigation projects. The remaining functions are to "promote" good ecological practices, and to "advise" on various strategies for water, irrigation and flood management.

The allocation and tariff-setting decisions have a wide impact on normative concerns of equity, accessibility and sustainability in particular, but the KWRA has a very vague mandate on allocations, and has been directed to ensure full cost recovery, leaving little room for any "independent" decision-making. Fixing entitlements is not under the current purview of the KWRA and neither is there an explicit mention of fulfilling the "right to water." There is also no mention of it being a conflict-and dispute-settling body either. The 21-member board is dominated by bureaucrats representing the Ministry of Water Resources, Water Resources Department, Rural Development and Panchayati Raj Department, Minor Irrigation Department, and the Finance Department–all handling IFI funded projects. While interviews with select members of the KWRA's board revealed a consistent lack of clarity on the roles and responsibilities of the newly constituted board, the general understanding is that the KWRA was created in an effort to bring together all water-related offices on the same platform to aid, what the government expects will be, a more holistic approach to issues in the water sector. Thus, the KWRA finally recognizes the hydrological connectivity between surface and ground water resources in the larger ecological landscape; whether such integration in major, medium and minor irrigation is really meant, is unclear. In future, it might be the nodal committee from which government orders can be issued. This would, in effect, imply that the Water Resources Department is restructured by centralizing decisions taken on water usage across sectors, such as for domestic purposes, irrigation, groundwater, and industrial usage, should the authority be given legal backing. The focus of the KWRA is the large and medium irrigation sector–nothing refreshingly new from what the Water Resources Department has been doing all along. Water security or even water utility services provisions, such as rural and urban drinking water, is not on its prime agenda. There is even less clarity on bringing about legislative changes to empower the water users associations and providing space for community participation in head works and main canal management that marred the Participatory Irrigation Management (PIM) initiatives in the state. Co-existence of regulatory (centralizing) institutions, such as the IRA alongside the PIM-type of initiatives that were introduced to bring in decentralization (deregulation), is a major institutional dilemma. Equity considerations are lacking in terms of water accessibility, affordability and distribution; instead, it has been understood in terms of providing for rehabilitation and resettlement. The KWRA has not yet dealt with issues of tariff setting; interviews with its members reveal that though they are not ideologically opposed to cost-recovery through tariff revisions, they would be reluctant to adopt higher tariff scales, at least until the people of Karnataka are "educated" at near-campaign levels, to co-operate with the government and pay for water as people have traditionally come to expect the delivery of water services through government subsidization. The creation of water markets and the trading of water entitlements, and promoting private players in service provision, is the vision. State control is seen as a return to the licence Raj, and its presence is deemed necessary only for interventions on the grounds of repercussions of exploitation of groundwater and water conflicts.

Realities of Reforms and Restructuring

The Maharashtra Water Resources Regulatory Authority (MWRRA), set up in 2005, was one of the first IRAs to be established (in India) in the state of Maharashtra, with Arunachal Pradesh following suit in 2006. The Uttar Pradesh Water Management and Regulatory Commission (UPWMRC) Act, 2008, was passed in the legislative assembly while Madhya Pradesh and Andhra Pradesh are awaiting pending bills along similar lines. The following section examines some of the provisions of the MWRRA and the UPWMRC with respect to crucial governance functions, and Karnataka's standing in some of these issues.

Water Pricing and Tariff Structures

The MWRRA Act (GoM 2005) and UPWMRC Act (GoUP 2008) empower water regulatory authorities to establish a tariff system, based on the principle of "cost recovery," and to determine and regulate water tariffs. The MWRRA Act restricts the level of recovery to operation and maintenance (O&M) cost whereas the UPWMRC Act provides for recovery of a part of the capital costs (in the form of depreciation) and "cost of subsidy" from the water tariffs along with O&M costs. Provision of recovery of capital costs paves the way for higher commercialization of water services and creates a favourable environment for privatization in the sector. The WB reportedly laid out six levels of cost recovery, and recommendations were made to achieve each higher level in February 2005 (Revels 2005). The lowest level of cost recovery would not even meet O&M costs, while the highest level would recover investments and yield profits, in addition.

Karnataka has been using increasing block tariffs, where billing is supposed to be proportionate to the degree of water consumption. While cross-subsidies are to be taking place between progressive tariff blocks, in his study, G. S. Sastry observes that "frequent tariff revisions during 1991–2002 have affected the lower consumption slabs, thereby nullifying the subsidy benefits enjoyed by them" (2004: 2). Also, poorer sections of society live in underserviced areas, and are forced to buy water from private tanker operators at higher prices. Thus, while increasing block tariffs are supposed to distinguish between providing water services as a social and as an economic good, the basic tariff structure in operation in

Karnataka itself does not successfully make that distinction, and it gets blurred further as other complexities of the quality of service compound it. Fundamental objections to rights-based obligations with respect to water provision are: (*a*) the state does not provide a basic minimum quantity of water free of cost or at some affordable fee to its citizens; (*b*) water is essential to life, and it's not just the service, but water usage that is being billed; and (*c*) in turn, water is being distributed on an ability-to-pay basis, rather than a need-based assessment of the different uses that scarce water may be put to, especially disregarding the many smaller livelihoods that depend upon water availability (Sangameswaran et al. 2008).

Water Entitlements and Allotments

Having an entitlement regime in India where citizens are assured of basic minimum water for drinking and livelihood needs would be a welcome step, as are priority-based allotments that ensure a basic minimum for all. Unfortunately, land rights have been a determining factor in water allocations, and efforts to delink them from water rights (though highlighted in the National Water Policy, 2002) are yet to be recognized.

Creation, management and regulation of "water entitlement system" (WES) is at the heart of the regulatory framework of the IRAs in the water sector. With this, various water users and groups of users shall be allotted certain shares of water as their "water entitlement." The UPWMRC and the MWRRA are empowered to determine and regulate water entitlements to different user groups. The UPWMRC Act defines entitlement as, "any authorization by the Commission to use the water for the specified purpose" (refer to Sect. 2 No (h) of UPWMRC Act) (GoUP 2008: 2). The MWRRA Act further states that, "entitlements . . . are deemed to be usufructuary rights" (refer to Sect. 11 No (i) of MWRRA Act) (GoM 2005: 14). That is, it is the MWRRA's responsibility to develop a framework and fix the criteria for trading in water entitlements where the authority is to determine the distribution of various user entitlements which can then be "transferred, bartered, bought or sold on annual or seasonal basis within a market system" (MWRRA 2005: 89). These are certainly not ownership rights but "rights to use" (usufructuary rights). Thus, "entitlements" are legally recognized, registered, (near) perpetual, and regulated rights over use of water.

Water Resources Monitoring, Management and Planning

Integrated Water Resource Management (IWRM) is being promoted through a reform process where the river basin is the basic unit of planning. It includes different sectoral uses of water and environmental needs under different land uses along with institutions that regulate the use of water. The concept of IWRM is adapted as Integrated State Water Plan (ISWP), at the state level, by water regulatory authorities. The UPWMRC Act allows the government to draft the ISWP and contains the powers of approval within itself. The MWRRA Act, however, accords the power of approval of the ISWP to a committee comprising various ministers, while the role of the MWRRA is limited to monitoring the ISWP's implementation. River Basin Agencies (RBA) created for the five river basins of Maharashtra are to prepare basin and sub-basin level plans with a multi-sectoral, participatory approach. The RBA plans are then integrated into the ISWP; however, issues like entitlements, tariffs, and prioritizing available water for equitable distribution are reserved for the MWRRA.

A decentralized river basin approach, highlighted in the MWRRA, to planning is indeed progressive. However, while the planning regime should ideally involve the government and citizens in the planning process, it is worrying that the IRA has the last word on the ISWP (as in UP), as it puts the planning regime under its direct control. Also, a clear concept of IWRM and a strategy of implementation have not been worked out between the local, state and national level. While the KWRA is mandated to adopt an integrated approach to water resource planning, using a river basin as the area of planning, it is not very clear what its role vis-à-vis the ISWP or RBA would be.

Private Sector Participation

Water privatization in Karnataka has several facets including the outright Private Sector Participation (PSP) model and its facelifted version (popularized by the state), the Public–Private Partnership (PPP) model; but, critically, the fundamentals of both models remain the same. The PSP was launched extravagantly in Karnataka, with initial pilots experimented in select wards of Hubli–Dharwad, Belgaum and Gulbarga. However, these experiments are now being considered for complete up-scaling to provide utility services to entire townships, with no significant changes in the modalities of the partnerships to include elements of public interest. Another example is an attempt to solve the water crisis in the peripheral areas of Bangalore, where the Government of Karnataka envisaged the Greater Bangalore Water Supply and Sewerage Project (GBWASP) to cover the eight ULBs (Dwivedi et al. 2007). The International Finance Corporation (IFC), of the World Bank group, has been assigned to make recommendations on appointment of a suitable private sector operator for the GBWASP. Core principles, such as beneficiary capital contributions, users' pay and full cost recovery, are being introduced in the GBWASP that will inevitably lead to denial of access to water for the urban poor. First, the lowest slab for one-time beneficiary contributions is INR 2,500 (US$ 55), which the urban poor cannot afford. Second, they will have to pay another INR 1,500 (US$ 38) for domestic water supply for individual houses. Third, it was made clear that there will be a hike in water tariffs. Those refusing to pay the beneficiary contribution will not be included in the scheme. In anticipation of this privatization of water supply, Bangalore Water Supply and Sewerage Board (BWSSB) has started to disconnect public stand posts in slum areas.

Private sector participation is well entrenched in Karnataka and the Manthan Database (Manthan 2009) can be referred to for exhaustive details on these projects and their status. Sangameswaran et al. (2008) also note that the experience (to date) with privatization in water supply services indicates that more critical engagement with the issues of need for 24/7, local institutional relations and equity is required before institutionalizing the process of privatization. Urs and Whittell (2009) situate water privatization and other development-linked water reforms with particular reference to Bangalore city.

Public Participation

Public–private partnerships are being heavily promoted by the IFIs in water sector reforms, having refrained from fully privatizing water utilities. Public participation (in these schemes) is largely confined to functional tasks of O&M and collection of fees. The IFIs have even outrageously termed the 10 percent contribution, in many rural water supply schemes, as public participation. Clearly, public participation is being solicited in a hollow manner by giving the public set of tasks to complete, which may contribute to a sense of ownership, but withholding decision-making powers. Most importantly, the public is not given a chance to say "No" to a project.

To include the public in decision-making, including resource allocation and distribution, and entitlements and allotments, where the ultimate decision-takers are accountable to concerns expressed by the public, gives them control over governance and, hence, is avoided. However, it also needs to be recognized that the "public" is often interpreted as a single class of the civil society. On the other hand, space has been provided for public representation (through Water Users Associations) in the composition of IRAs, but it is debatable whether it an eyewash, meant to justify water-sector reforms, or whether a genuine space has been created for concerns of the public to shape the decisions. For instance, while public representation has been sought in the KWRA, through representation from farmers' unions, water user associations and some NGOs, these representatives are severely outnumbered on the 21-member board (even on paper) with representatives from various unions yet to be identified. The minutes of the three meetings held by the authority so far showed no indication of an active attempt to fill in these crucial vacancies (GoK 2009). The skewed composition of the KWRA Board has resulted in few civil society members finding support for their views and contributions towards the working of the Board as they are in the minority.

Transparency, accountability and participation within the system lean more towards being investor friendly, rather than creating an environment for a socially responsible and/or accountable system (Wagle and Dixit 2006). Decision-making powers remain with institutions that are not easily accessed by the public, and where there is no process by which decision-makers can be held accountable for disregarding public views and suggestions in their final decisions.

Governing Options

India has conventionally followed the "state hydraulic paradigm of water management" where the state was the custodian of water resources. However, proponents of the "state failure hypothesis," seeking to privatize the water sector, argued that the state continues to remain unproductive, ineffective and inefficient, thereby ushering "market environmentalism" as the alternative model for water governance (see Megginson and Netter 2001; Kikeri and Nellis 2002; Kemp 2007). Market environmentalism typically ensures that water quality is maintained by accounting for environmental and conservation costs in the manner in which water is priced, in addition to other infrastructure and economic costs. Thus, pricing (or economic efficiency), rather than access or equity, is prioritized, often leading to private property rights and tradable markets for water (Bakker 2005). Government regulation also centers on market-based or market-stimulating techniques and the private sector begins to play an important role as the owner and manager of infrastructure and water delivery services. Though the state hydraulic model is being loosely replaced by the market environmentalism model for water governance in many states of India including Karnataka, many long-standing problems with water quality and service provisioning remain. Considerations of ecosystem sustainability are far from being addressed in the new models of governance.

The IRA, a vehicle of market environmentalism, cannot by definition or practise accommodate the principles that social organizations and environmental groups see as essential to water governance. The independent regulatory body is a new institution, which does little that the government is not capable of doing itself with proper public participation. While there is no doubt that the current post-reforms period is deeply problematic, the pre-reforms period was not desirable either. Gaps in sustainable water use, minimum ecosystem flows, cost-effectiveness, efficiency, and equitable distribution of water were not met in the pre-reforms period either; however, water sector reforms and the introduction of independent regulation makes the possibility of the state providing water for all, as a measure of social security, bleaker. At the core of the argument is whether or not privatization can provide the answers to issues where the State failed to provide; the little experience in the Indian context suggests that they cannot.

In Mexico, implementation of comprehensive water sector reforms started in the1990s and since then the government has managed to (*a*) create a new legal framework; (*b*) restructure existing water administration; (*c*) promote and support a plurality of new autonomous and quasi-autonomous water institutions; (*d*) modify incentives in water use to different user groups; and (*e*) struggle with a vast complex of unresolved operational issues in implementing the reforms (Shah et al. 2004). While reforms implemented in Mexico may not be replicable for India, their experience suggests commitment to the reform process that regards social justice and environmental sustainability.

There are also an increasing number of examples to show resistance to private operators in the field of water service provision (an established trend) described as the "remunicipalisation wave" (Water Remunicipalisation Tracker 2010). According to the Tracker, remunicipalisation is happening at municipal and community levels (such as in France or the US) and at regional levels (as in Buenos Aires and the Santa Fe provinces in Argentina) and at national levels (such as Uruguay and Mali). Around 40 municipalities and urban communities in France have already taken water services back into the public, over the last 10 years, resulting in cheaper tariffs and improved services. Cities in the US, large and small, have remunicipalised their water services as a reaction to poor service and excessive rates (ibid.). In both countries, some private operators used sophisticated and dishonest management and financial practices to increase profits. At the same time, innovative public water management reforms are taking place across the globe, as in Bolivia, Uruguay and Buenos Aires. Remunicipalisation campaigns are also underway in parts of Argentina, Mexico, Italy, and Northern Ireland. In keeping with the concept of remunicipalization, alternatives for regulation from within the government are also possible. Indeed, Hood and Scott (2000: 7) identify seven types of oversight organizations inside the government, constituting a "web of regulation over government," such as: (*a*) international bodies like the World Trade Organization (WTO); (*b*) bureaucratic agents of legislatures, like the National Audit Office in the UK; (*c*) grievance-handlers, such as ombudsmen, outside the normal public law framework; (*d*) bodies independent of both legislature and executive, but with public responsibilities, such as committees on merit in public appointments or standards of conduct in public life; (*e*) quasi-independent executive bodies, such as inspectorates; (*f*) bodies that regulate across public and private sectors, such as regulation of health and safety at work; and (*g*) independent private bodies with the ability to affect states, for example, credit rating organizations.

A parallel concept, Public–Public Partnership (PUP) is also gaining wide acceptance in certain countries within Latin America and the Asia-Pacific. PUPs are new ways in which partnerships are developed between public water operators, communities, trade unions and other stakeholder groups, without profit motive and on a basis of equality (Corral 2007). In addition, PUPs are also being created when well-performing public utilities are matched or twinned with those that are not-so-well performing to share expertise on a not-for-profit basis and improve the standard of the lesser performing utility. V. P. Corral (2007) suggests that PUPs are most effective when all partners have an understanding of each other's goals and are willing to work together to reach their shared goals, while addressing the common causes of public service failure, and secure affordable clean water for all.

India's dissatisfying experiences with IFI-initiated reforms in the water sector may well be the impetus for it to also explore options of "remunicipalization" and PUP, while restructuring the partnering arrangements in private sector partnerships. However, this would mean making adequate provisions in the current legal frameworks to represent and prioritize public interest.

Conclusion

The challenge of water governance is to be informed by certain principles based on equity, sustainability and public participation. As a custodian of water resources, the government needs to circumscribe water governance in the context of "Right to Water," which can be understood in two ways—as a basic right stemming from human dignity and as an aspect of legal recognition in international and national law. Issues surrounding water rights are of importance to social justice and need concrete guiding principles, as power equations between the various sections of society and resource users could easily violate these principles. Aspects of accessibility, affordability, ownership, equity, equitable distribution, delivery, and participation constitute some elements of the "Right to Water" that should be based on democratic decisions.

Thus, "public control over governance", a useful concept that requires people-centered transparency, accountability, participation, and autonomy, also needs to be introduced so that the public has legal rights to intervene at any point of the process when it is felt that the government or implementing/governing agencies are deviating with results that go against the public interest. Any water governance system, in Karnataka or elsewhere, should be premised upon principles (such as these) which are socially and environmentally sound. But the IRA as an institution (by definition) and in practise cannot accommodate these basic principles. While inadequate work has gone into the understanding of how IRAs could be remodelled to best suit a country like India (and for specific conditions, such as in Karnataka), the basic question of whether independent regulators are indeed the appropriate choice for water governance has not been addressed. So far, it has only been assumed and imported from other contexts without addressing the need for reforms that affect larger issues of environmental sustainability and social concerns.

References

Bakker, K. 2005. "Neoliberalizing Nature? Market Environmentalism in Water Supply in England and Wales," *Annals of the Association of American Geographers*, 95(3): 542–65.

Barlow, M. and T. Clarke. 2002. *Blue Gold: The Fight to Stop the Corporate Theft of the World's Water*. New York: New Press.

Corral, V. P. 2007. "Public–Public Partnerships in the Water Sector." Paper presented at "World Water Challenges and Japanese Standpoints," Open Event at the First Asia–Pacific Water Summit, co-organized by JTUC-RENGO and PSI-JC (PSI-Japan Committee), December 2, Oita, Japan. Available at http://www.waterdialogues.org/documents/Public-PrivatePar tnershipsintheWaterSector.pdf (accessed December 25, 2010).

Dubash, N. K. 2008. "Independent Regulatory Agencies: A Theoretical Review With Reference To Electricity and Water in India," *Economic and Political Weekly*, 43(40): 43–54.

Dwivedi, G., Rehmat and S. Dharmadhikary. 2007. *Water: Private, Limited: Issues in Privatisation, Corporatisation and Commercialisation of Water Sector in India*. Badwani: Manthan Adhyayan Kendra. Available at www. manthan-india.org/IMG/pdf/Water_Pvt_Ltd_New.pdf (accessed March 10, 2010).

Government of India (GoI). 2002. *National Water Policy*. Available at http://mowr.gov.in/writereaddata/linkimages/nwp20025617515534.pdf (accessed March 10, 2010).

Government of Karnataka (GoK). 1965. *The Karnataka Irrigation Act*. Available at http://www.karnataka.gov.in/minorirrigation/pdf/irrigation_act.pdf. (accessed March 10, 2010).

———. 1999. *The Karnataka Ground Water (Regulation for Protection of Sources of Drinking Water) Act*. Available at http://www.cseindia.org/userfiles/ KarnatakaGWact.pdf (accessed March 10, 2010).

———. 2002. *Karnataka Urban Drinking Water and Sanitation Policy*. Available at http://www.ielrc.org/content/e0332.pdf (accessed March 10, 2010).

————. 2002. *State Water Policy 2002*, edited by W. R. Department. Available at http://waterresources.kar.nic.in/state_water_policy-2002. htm (accessed March 10, 2010).

————. 2004. *Karnataka Municipal Corporations (Water Supply) Rules.* Available at www.indiawaterportal.org/sites/indiawaterportal.org/files/ e0404.pdf (accessed March 10, 2010).

————. 2006. *The Karnataka Groundwater (Regulation and Control of Development and Management) Bill.* Available at http://dpal.kar.nic.in/ pdf_files/.%5C25of2011(E).pdf (accessed March 10, 2010).

————. 2008. Government Order No. WRD 85 MDI2008 Bangalore Dated 08.09.2008, edited by W. R. Department. Government of Karnataka.

————. 2009. Agenda Notes of the Water Resources Authority Meeting. Water Resources Department, Vikasa Soudha, Karnataka. Unpublished notes accessed via personal communications.

Government of Maharashtra (GoM). 2005. *Maharashtra Water Resources Regulatory Authority Act.* Available at http://mahawrd.org/downloads/ MWRRA.pdf (accessed March 10, 2010)

Government of Uttar Pradesh (GoUP). 2008. *The Uttar Pradesh Water Management and Regulatory Commission Act.* Available at http://www. cseindia.org/userfiles/UP_water_mgmnt_reg_act08.pdf (accessed March 10, 2010)

Grusky, S. 2001. "Privatization Tidal Wave IMF/World Bank Water Policies and the Price Paid by the Poor," in *Bearing the Burden of IMF and World Bank Policies.* Multinational Monitor, September 2001, 22(9). Available at http:// www.multinationalmonitor.org/mm2001/01september/sep01corp2.html (accessed March 25, 2010).

Hood, C. and C. Scott. 2000. "Regulating Government in a 'Managerial' Age: Towards a Cross-National Perspective." Discussion Paper 1, Centre for Analysis of Risk and Regulation, London School of Economics and Political Science, London. Available at http://www2.lse.ac.uk/ researchAndExpertise/units/CARR/pdf/DPs/Disspaper1.pdf (accessed March 10, 2010).

Kemp, R. L. 2007. *Privatization: The Provision of Public Services by the Private Sector.* Jefferson, NC: McFarland & Company Publishers.

Kikeri, S. and J. Nellis. 2002. "Privatization in Competitive Sectors: The Record to Date." World Bank Policy Research Working Paper 2860, World Bank, Washington, DC.

Karnataka Urban Infrastructure Development and Finance Corporation (KUIDFC). 2007. *Establishment and Operationalization of Karnataka State Urban Water Supply Council: Final Business Plan.*

Manthan. 2009. "Manthan Databases of Privatization and Reforms Projects." Available at http://www.manthan-india.org/spip.php?article23 (accessed October 15, 2009).

Megginson, W. L. and J. M. Netter. 2001. "From State To Market: A Survey of Empirical Studies on Privatization," *Journal of Economic Literature*, 39(2): 321–89.

Moench, M., A. Dixit, S. Janakarajan, M. Rathore, and S. Mudrakartha. 2003. *The Fluid Mosaic: Water Governance in the Context of Variability, Uncertainty, and Change.* Kathmandu: Nepal Water Conservation Foundation. Available at www.i-s-e-t.org/images/pdfs/Fluid%20Mosaic%2003.pdf (accessed March 10, 2010).

Ramchandrudu, M. V., S. Warghade, and S. Wagle. n.d. "Emerging Water Regulatory Frameworks and Related Reforms in Instruments of Water Governance in Andhra Pradesh: Discussion Note for Facilitating Grounded Analysis and Public Debate." Unpublished report accessed via personal communication.

Revels, C. 2005. "'Equitable' Cost Recovery: Do We All Mean the Same Thing?" Paper presented at "Water Week 2005." Available at http://siteresources.worldbank.org/EXTWAT/Resources/4602122-1213366294492/5106220-1213804320899/7.1Equitable_Cost_Recovery.pdf (accessed March 10, 2010).

Rogers, P. and A. W. Hall. 2003. "Effective Water Governance." TEC Background Paper No. 7, Global Water Partnership, Stockholm.

Sangameswaran, P., R. Madhav and C. D'Rozario. 2008. "24/7, 'Privatization' and Water Reform: Insights from Hubli–Dharwad," *Economic and Political Weekly*, 43(14): 60–67.

Sangameswaran, P. 2009. "Neoliberalism and Water Reforms in Western India: Commercialization, Self-Sufficiency, and Regulatory Bodies," *Geoforum*, 40(2): 228–38.

Sastry, G. S. 2004. "Urban Water Supply and Demand: A Case Study of Bangalore City." ISEC Working Paper Series, Ecological Economics Unit, Bangalore.

Shah, T., C. Scott and S. Buechler. 2004. "Water Sector Reforms in Mexico: Lessons for India's New Water Policy," *Economic and Political Weekly*, 39(4): 361–70.

Urs, K. and R. Whittell. 2009. *Resisting Reform? Water Profits and Democracy.* New Delhi: Sage Publications.

Wagle, Subodh. 2005. "Revisiting Good Governance: Asserting Citizens' Participation and Politics in Public Services," in Daniel Chavez (ed.), *Beyond the Market: The Future of Public Services.* London: Transnational Institute, Amsterdam and Public Services International Research Unit.

Water Remunicipalisation Tracker. 2010. Available at http://www.remunicipalisation.org/ (accessed March 10, 2010).

World Bank. 2008. *Country Strategy for the Republic of India for the period FY2009–2012.* International Bank for Reconstruction and Development, International Development Association, International Finance Corporation, World Bank.

PART III

Urbanization and Water: Emerging Conflicts, Responses and Challenges for Governance

11

Urbanization and Water

A Conundrum and Source of Conflict?

Vishal Narain

A radical demographic shift has occurred worldwide, characterized by the movement of people from urban to rural areas at an increasing rate. While in the mid-1970s less than 40 percent of the world's population lived in urban areas, it has been estimated that by 2025, 60 percent of this population will do the same (World Bank 2000). The world's urban population today is around three billion people, about the same size as the world's total population in 1960 (Satterthwaite 2006). Around 50 percent of the world's population now lives in urban centers, compared to less than 15 percent at the beginning of the century.

Urbanization in Developing Countries

Two striking features of this process of urbanization are noted. First, there is a large concentration of urban population in developing countries. It is believed that changes in the urban population will particularly affect low income countries in the future (World Bank 2000). In 1950, 41 of the world's 100 largest cities were in developing countries. By 1995, the number had risen to 64 and has been growing exponentially eversince. The urban population of Africa, Asia, Latin America, and the Caribbean is now nearly three times the size of the urban population of the rest of the world (World Bank 2009). Projections of the United Nations had suggested that that 85 percent of the growth in the world's population between 2000 and 2010 would be in the urban areas, especially in Africa, Asia and Latin America. Cities in developing countries are expected to double in three decades, adding another two billion people to their population (World Bank 2009).

Second, though concerns regarding this rapid urbanization tend to focus on large cities, half of the world's urban population and a quarter of the entire population lives in urban centers with less than half-a-million inhabitants (Satterthwaite 2006). David Satterthwaite (2006) notes that the increasing number of mega-cities with 10 million or more inhabitants may seem to be a cause of worry or concern but there are relatively few of them across the world; in 2000, there were 18 of them. They contained less than 5 percent of the world's population and were heavily concentrated in the world's largest economies. On the other hand, if we take small urban centers to mean all urban settlements, as defined by governments, with a population of less than half-a-million inhabitants, then by 2000, about 1.5 billion people lived in small urban centers, including more than a billion in low-and middle-income countries. Once again, nearly half of this population lived in Asia; and nearly a quarter in Europe.

The Drivers of Urbanization

Urbanization is considered to be an irreversible force driven, at least in part, by an economic shift in many countries from agriculture to industry, trade and services (World Bank 2000). Spatial transformations–the growth of cities and leading areas–are linked closely to changes in the economy, especially the sectoral transformations that accompany growth and the opening of an economy to foreign trade and investment (World Bank 2009). Further, urbanization processes have been understood to be propelled by the forces of globalization, a subject which is widely contested, and which has been discussed in the opening chapter of this book.

With the onset of urbanization processes globally, social scientists and urban theorists started devoting much of their intellectual energies to understanding the growth of cities. It came to be widely understood that cities emerge and grow when increasing returns exceed transportation costs and that spatial concentration itself creates a favorable economic environment by providing technological spillovers and economies of scale. In *The Economy of Cities* (1969), Jane Jacobs developed a theory in which early cities were understood to support their surrounding areas, disputing the notion that cities build upon a rural economic base. Social theorists of the 19th and early 20th century equated increasing urbanization with progress (Bienen 1984). Development economists, looking at Asia, Africa and Latin America after World War II, considered urbanization to be a positive development. Rising standards of living were associated with growth of cities in Western Europe and the United States. Cities had higher productivity and social services could reach a larger number of people; new technologies were also generated and disseminated. In general, urbanization came to be equated with progress.

The Ecological Foot-print of Growing Cities

Cities are characterized by several changes, in particular an ever-increasing metabolism and ecological foot-print as reflected in the growing demand for natural resources like land and water. Cities are in a state of transition, distinguished by a wide range of changes and spatial divisions (Marcuse and van Kempen 2000). Like other assemblies of organisms, cities have a definable metabolism, consisting of the flow of resources and products through the urban system, for the benefit of urban populations (Gerardet 2004). This metabolism is understood to be essentially linear, with resources being pumped through the urban system without much concern about their origins or the destination of wastes, resulting in the discharge of vast amounts of waste products incompatible with natural systems.

The implications of urban expansion for the use of natural resources are perhaps best understood using the concept of ecological foot-print. Ecological foot-print is an accounting tool that enables us to estimate the resource consumption and waste assimilation requirement of a definite human population or economy in terms of a corresponding productive land area (Wackernagel and Rees 2004). The total ecosystem area that is essential to the continued existence of the city is considered to be its *de facto* ecological foot-print on the earth. It is proportional to both population and per capita consumption. For modern industrial cities, the area involved is considered to be of the order of magnitude larger than the area physically occupied by the city; it represents the corresponding population's "appropriated carrying capacity." It is estimated, for instance, that for urbanization, food, forest products, and fossil fuel use, the Dutch consume the ecological functions of a land area over 15 times larger than their country (Wackernagel and Rees 2004).

The concept of ecological foot-print, thus, gives us some answers to the questions or estimates regarding a city's resource metabolism (Wackernagel et al. 2006). By measuring the overall supply of and human demand on regenerative capacity, the ecological foot-print serves as an ideal tool for tracking progress, setting targets and driving policies for sustainability. The concept has been important in encouraging urban planners and environmental managers to look beyond the traditional scales of planning and management to consider the regional and international environmental impacts of a city's activities (McManus and Haughton 2006).

It is now widely understood that the ecological foot-print of cities spills well over their geographical limits, which can be an important source of conflict with the peripheral areas–as demonstrated by all six chapters in Part 3 of this book. The analyses show how urban settlements depend on their hinterlands as (*a*) a source of natural resources and rural products, (*b*) as sinks for wastes, and (*c*) as sites for expansion. Urban expansion transforms the use patterns of the land–whose use is determined by the demand for both land-based products and for resources, such as water, whose appropriation changes land-use patterns (McGranahan 2006).

The South Asian Context: Conflicts and Contestations over Scarce Water

According to Ruth Meinzen-Dick (2000), water is understood to have many values–social, ecological, religious, cultural, and political. She suggests that a closer look at the actual usage of municipal water reveals that very little, in fact, of it goes into drinking, since industrial, civic and business consume a large part of it. Urban and peri-urban agriculture may also use some portion of municipal water supply systems, either in small home gardens or in commercial horticultural systems and production centers in adjoining cities.

As the demand for the water in several South Asian cities keep growing, these cities look further and further for their water sources;

therefore, the phenomenon of acquiring water from other uses, notably agriculture, in cities like Kathmandu, Ahmedabad and Chennai, as also smaller towns and urban centers, has become common (Meinzen-Dick 2000). Ruth Meinzen-Dick notes that such water transfers may be private, unplanned and ad hoc, with individual well owners pumping water into tankers to be sold in the city, or public and planned, with water districts taking water from the villages for selling to the city, with or without compensating them. This gives rise to several issues of equity; is it fair for industry and urban centers to use the water, which has conventionally been used by rural inhabitants for purposes of agriculture or domestic consumption?

Growing competition between rural and urban presents ripe opportunities for conflict, and many studies in the region have begun to highlight that. For instance, S. Janakarajan (2009) describes the escalation of conflicts, regarding Chennai, at the rural-urban level; he notes that panchayats have been sidelined and a threshold level of scarcity is needed to mobilize the peri-urban residents. A similar observation is made for Gurgaon (Narain 2009a). In the third week of March 2008, farmers living near Gurgaon breached the Gurgaon canal, which is the major supplier of water, forcing the residents to buy water from private sources (tankers) at prices as high as INR 500–700 per 5,000 liters. About 400 water tankers had to be pressed into service to supply tube well water to the people of Gurgaon on March 24, 2008, which could only meet just about 30 percent of the total demand for water. This water crisis was seen as an outcome of the short-sightedness of the government in issuing licenses for malls and residential areas without taking cognizance of the water availability.

Urbanization affects the access of peri-urban residents to water in several ways (Narain 2009a) creating the potential for conflict. The residents are often left chasing the water table, as pressures on groundwater increase from water-guzzling factories and farm houses. It is not simply that they lose access to water as it is physically transported to the cities, but also that water resources in their own locations may be pre-empted by the resource rich, who are able to afford extraction from the depths of the aquifers. Urbanization processes often bring the urban elite into the peripheries, looking for cheap land or other avenues, for investing their wealth. They can afford costly water extraction technologies, depriving the locals of access to this resource. For instance, in one of the villages (where the author is currently studying peri-urban water use dynamics), local residents have begun chasing the water table as farm houses, a major "rural" land use of the "urban" elite, have pre-empted the groundwater, using submersible pump-sets, digging much deeper than the local residents. When the groundwater underlying their farm houses was saline, they bought small parcels of land overlying fresh groundwater and transported it through underground pipes to their farm houses over distances of 2–3 kilometers.

The usual articulation of the rural-urban water nexus is in terms of the physical movement of water from rural to urban areas, as demonstrated, for instance, in the transportation of water from villages to urban centers through water tankers. However, to understand the implications of urbanization, one needs to look at the wider variety of ways in which urbanization affects rural water use, rather than the sheer transfer of water alone. As cities grow and urban populations multiply, urban authorities typically respond to supply augmentation by creating additional water supply infrastructure. The supply of water to cities involves the development and building of water treatment plants that are usually built on land acquired from the peripheral areas. When peri-urban residents lose this land, they also lose the access to water sources located on those lands. In other words, peri-urban residents lose both land and water to provide water to the growing cities. Similarly, when water needs to be transported from distant sources, it is done through the canals and channels that pass through the peripheral villages and for which lands are acquired from such areas.

Their access further diminishes as common lands, on which local water sources such as village ponds are located, are acquired for industrial and urban residential purposes (Narain 2009b). Their routes to water sources get disturbed as they are dissected by lands acquired for the construction of highways. As they diversify occupationally and migrate to the cities, the level of interest in maintaining local water sources and other common property resources (CPRs) tends to diminish. The location of sewage treatment plants (to supply water to the city) disturbs the local quality of life, often raising the water table level, culminating in water-logging and a gradual loss of agricultural productivity (Narain 2009b). At the same time, the relocation of factories from the city to the peripheries also contaminates groundwater aquifers (Narain and Nischal 2007). Sewage-based agriculture has emerged as an important form of urban–rural resource flows in South Asia–notably in India and Pakistan–but only a few farmers may be able to benefit from it, depending on the location of their fields (Narain 2009b). Further, sewage-irrigated crops adversely affect the health of irrigators as well as the consumers of the resultant crops.

The pre-emption of scarce groundwater supplies for industrial use, such as for manufacturing beverages, brought popular attention to the subject of use and ownership rights over water (Drew 2008). This has taken the form of a debate on public vs private ownership of groundwater, as in the case of Plachimada, Kerala, India, wherein the Perumatty panchayat chose to support the tribal women, who were conducting an infinite sit-in, to stop the loss of an estimated 1.5 million liters of water, a day. After filing a Public Interest Litigation (PIL) in the Kerala High Court, the Chief Minister ordered the closure of the beverage plant on February 17, 2004. The panchayat continued to withhold permission to the company to resume operations (Ranjith 2004). Such situations point to a conundrum regarding the use of water between competing rural and urban uses.

Exploring the Conundrum: The Land–water Nexus and Property Rights Issues

Why is it a conundrum? In large part, it is a conundrum because there is a question of who owns the water. This could be seen as the outcome of a situation in which access to to water is tied to ownership of land. Thus, as lands are acquired–and this has been the usual mechanism for the expansion of cities–the owners of those lands *de facto* lose access to sources of water to which they had access because they had previously owned those lands. The current debate in the media on the subject has centered on financial compensations to landowners for land acquisition, but in the absence of a property rights structure for water, the diminishing access of peri-urban residents to water has received little attention (Narain 2009a).

The following six chapters demonstrate the diversity of ways in which urbanization processes escalate rural–urban water conflicts, both in terms of the quantity and quality of water. These present a very good overview of not only how urbanization processes have influenced the demand for water in the region, but also (and perhaps

202 ～ *Vishal Narain*

more importantly) how they have shaped and impacted upon water management practices, institutions and technologies.

Bijaya and Sushmita Shrestha's analysis shows how urbanization processes, shaped by neoliberal policies, have eroded the fit between technology, society and settlement in the Kathmandu Valley. The two chapters from India–from the southern state of Karnataka and the western state of Gujarat–describe the various consequences and conflicts that emerge, as pressures from industry and other urban uses over the same resource base increase. The contamination of water by industry adversely affects the productivity of agriculture and fisheries. At the same time, there are important equity issues; within the "rural" sector, the poor end up paying much more when they have to buy water from unorganized sources in the face of diminishing access to organized sources of water supply.

The other two chapters–from Bangladesh and Sri Lanka, respectively–make a closer link with governance issues. The chapter from Bangladesh reveals the contamination of the rivers around the city of Dhaka to be an outcome of several weaknesses in the governance structure; the chapter from Sri Lanka reveals how in the context of globalization, a greater role for donors and funders has created a pattern of donor–dependence among water utilities in Kandy–it also suggests a way out in terms of the approaches and interventions that may be needed to manage water more effectively.

All six chapters agree and imply that the water scarcity (that seems to trigger conflicts) is much less a physical phenomenon than an institutional one. To that extent, while they do paint a dismal picture of what is happening on the ground, they also give us hope in pointing out that correcting certain institutional weaknesses can ameliorate the situation to a large extent. That is, the conundrum and conflicts are not inevitable if water can be managed more effectively, through important changes in the institutional framework for water management.

References

Bienen, H. 1984. "Urbanization and Third World Stability," *World Development*, 12(7): 661–91.
Drew, G. 2008. "From the Groundwater Up: Asserting Water Rights in India," *Development*, 5(1): 37–41.
Gerardet, M. 2004. "The Metabolism of Cities," in S. M. Wheeler and

T. Beatley (eds), *The Sustainable Urban Development Reader,* 2nd edition, pp. 1157–64. Urban Reader Series. London; New York: Routledge.

Janakarajan, S. 2009. "Urbanization and Peri-urbanization: Aggressive Competition and Unresolved Conflicts–The Case of Chennai City in India," *South Asian Water Studies,* 1(1): 51–76.

Marcuse, P. and R. van Kempen. 2000. "Introduction," in P. Marcuse and R. van Kempen (eds), *Globalizing Cities: A New Spatial Order,* pp. 1–21. Oxford; Malden: Blackwell Publishers.

McGranahan, G. 2006. "An Overview of Urban Environmental Burdens at Three Scales: Intra-Urban, Urban-Regional and Global," in C. Tacoli (ed.), *The Earthscan Reader in Rural–Urban Linkages,* pp. 298–319. London; Sterling: Earthscan.

McManus, P. and G. Haughton. 2006. "Planning with Ecological Foot-Prints: A Sympathetic Critique of Theory and Practice," *Environment and Urbanization,* 18(1): 113–27.

Meinzen-Dick, R. S. 2000. "Values, Multiple Uses and Competing Demands for Water in Peri-urban Contexts," *Water Nepal,* 7(2): 9–12.

Narain, V and S. Nischal. 2007. "The Peri-urban Interface in Shahpur Khurd and Karnera," *Environment and Urbanization,* 19(1): 261–73.

Narain, V. 2009a. "Gone Land, Gone Water: Crossing Fluid Boundaries in Peri-urban Gurgaon and Faridabad, India," *South Asian Water Studies* 1(2): 143–58.

———. 2009b. "Growing City, Shrinking Hinterland: Land Acquisition, Transition and Conflict in Periruban Gurgaon, India," *Environment and Urbanization,* 27(2): 501–12.

Ranjit, K. R. 2004. *Holy Waters from the West.* Thrissur: Altermedia.

Satterthwaite, D. 2006. "Small Urban Centers and Large Villages: The Habitat for Much of the World's Low Income Population," in C. Tacoli (ed.), *The Earthscan Reader in Rural–Urban Linkages,* pp. 15–38. London; Sterling: Earthscan.

———. 2008. "The Social and Political Basis for Citizen Action on Urban Poverty Reduction," *Environment and Urbanization,* 20(2): 307–18.

Wackernagel, M. and W. Rees. 2004. "What is an Ecological Foot-print?," in S. M. Wheeler and T. Beatley (eds), *The Sustainable Urban Development Reader,* 2nd edition, pp. 91–98. Urban Reader Series. London; New York: Routledge.

Wackernagel, M., J. Kitzes, D. Morgan, S. Goldfinger, and M. Thomas. 2006. "The Ecological Foot-Print of Cities and Regions: Comparing Resource Availability with Resource Demand," *Environment & Urbanization,* 18(1): 103–12.

World Bank. 2000. *World Development Report 1999–2000.* Washington, DC: World Bank.

———. 2009. *World Development Report 2009: Spatial Disparities and Development Policy.* Washington, DC: World Bank.

12

Contextualizing Rural–Urban Water Conflicts

Bio-Physical and Socio-Institutional Issues of Domestic Water Scarcity

Shrinivas Badiger, Smitha Gopalakrishnan and Iswaragouda Patil

━━

Increasing demands for domestic water from agriculture and industrial sectors are causing catastrophic magnitudes of water scarcity in smaller urbanizing towns, where there is a direct and close conflict (even inter-sectorally) over dwindling water resources, making water management and domestic water supply a major public utility issue and governance challenge. The agriculture sector, mainly in the form of irrigation, consumes 85 percent of the annual utilizable water. The estimates by the Ministry of Water Resources (MoWR) (GoI 2006) indicate that by the year 2050, India will require five times more water for industries to maintain anticipated rates of economic growth, while the irrigation water demand will rise by 50 percent to meet the food productions needs, and drinking water demand would double. One of the major externalities of increasing agricultural and industrial water demands, apart from depleting environmental resource base, is its unintended impacts on drinking water sources. This has resulted in increasing number of conflicts between small urbanizing towns, villages and irrigation users of river and groundwater systems.

Urban population in India will reach around 40 percent of the country's population by 2020, and by 2025, more than 50 percent of the population will live in cities and towns (GoI 2002a). Along with rapid urbanization, there is also a trend towards the concentration of urban populations in smaller towns, rapidly pushing these small urban settlements up the next tier of recommended classes for cities and urban agglomerations. The percent share of the lower order class of towns (Class IV, V and VI), with less than 20,000 people, dropped from 47 percent (in 1901) to 10 percent (in 1991). An important distinction between the higher and middle order towns is their supporting economic activities. Scholars have argued that the Class I towns/cities experience relatively high and stable demographic growth because they are linked to the national and, sometimes, global markets (Kundu et al. 1999). Such integration with the higher order of the economic system is largely absent in the smaller towns (Class II and III) where the primary economic drivers are agriculture and its subsidiary occupations. In smaller towns, which are mostly rooted in their regional economy, however, population growth tends to be lower and fluctuates over time and space. This puts tremendous pressure on Urban Local Bodies (ULBs) in smaller towns to improve water infrastructure and supply services to meet the rising domestic water demands, without the kind of financial support available for Class I towns.

Popular notions of water scarcity, within the policy and academia, have often projected it as the result of the finite nature of water resources within competing production environments (Shiklomanov 1998; Falkenmark and Widstrand 1992). Recent methodologies have also used political ecology, common property resource theory and post-institutional approaches to highlight that scarcity is more than a natural condition (Mehta 2003; Mehta 2007; Ohlsson and Turton 2000). This chapter uses (*a*) conceptual approaches of the physical sciences to describe issues concerning physical resource shortage, and (*b*) socio-institutional analysis to describe the systemic causes of scarcity. The case presented here is of the town of Bailhongal, located in the Malaprabha reservoir catchment, in the northern regions of semi-arid Karnataka, India. The chapter begins with an assessment of the physical resources and elaborates on the various bio-physical factors in the town that add up to felt water shortage in the town and the conflict situation at the catchment and local scales. It concludes by evaluating the supply system components, among the various operational factors and takes an overview of the policy and institutional mechanisms that contribute to the construction of water scarcity.

Status of Supply and Scarcity

Malaprabha River plays an important role in the overall economy in parts of Belgaum and Dharwad districts of Karnataka. While the river is a major source of domestic water and irrigation supply for several villages and small towns, the reservoir at Saundatti is also a primary source of domestic water supply for Hubli and Dharwad. Additionally, the reservoir provides irrigation for several villages in the Belgaum, Dharwad and Bagalkot districts through its vast canal infrastructure connecting the right and left bank canals. Bailhongal is a *taluka* headquarter located in the reservoir catchment with a population density of 4,350 people/sq. km, which has nearly doubled reaching a population of 47,000 during the period 2001–10. In this section, we analyze the construction of felt water shortage in Bailhongal, described as both "natural" and "manufactured," by identifying factors associated with water provisioning. Apart from secondary data, focus group discussions (FGDs), semi-structured household surveys and interviews with elected members of the town municipal council and officials of the Water Supply Department (conducted during 2007–8) support the inferences in this section.

Water Resources Availability

Malaprabha reservoir catchment (see Map 12.1) lies in the semi-arid agro-climatic zone that receives an average annual rainfall of 625 mm. Though there have been no perceivable changes in the net annual rainfall over the last three decades, our analysis of the rainfall times series suggests that there are signals in the rainfall pattern that indicate decrease in the number of rainy days coupled with higher daily rainfall intensities. The traditional sources of domestic water supply in Bailhongal town were the village tank, for washing and other domestic needs and the open wells, constructed around the tank, for drinking water purposes. Apart from storing the runoff, the tank also served a critical function of recharging the shallow and

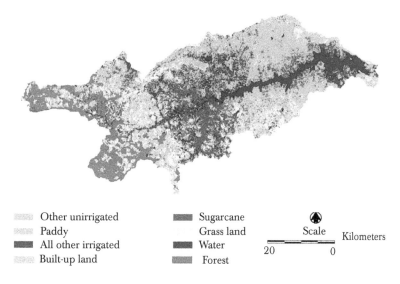

Other unirrigated	Sugarcane	⬥	
Paddy	Grass land	Scale	Kilometers
All other irrigated	Water	20　　　　0	
Built-up land	Forest		

Map 12.1: Land use composite in Malaprabha catchment and
location of Bailhongal town

Source: IRS-P6 LISS3 Satellite Images (2007).

deep aquifers, tapped by wells during the entire year. Since 1976, the town has been withdrawing water from the Malaprabha River, close to the backwaters of the Malaprabha reservoir, situated seven kilometers south of the town.

The municipal water supply infrastructure of Bailhongal was designed for a supply capacity of 90 MLD (million liters per day) equivalent to 2,000 lpcd (liters per capita per day), for a population of 25,000 in 2003, or approximately 1,000 lpcd for the current population estimate of 47,000 at 50 percent supply efficiencies. However, the perceptions of the community in the old town reveal that the supply levels were adequate until the 1990s, and are far below 1,000 lpcd. Open wells have also become less reliable sources of water during summer and drought years. Instances of bore well failure have increased in the town, since the last decade, suggesting that extraction from deep aquifers is far exceeding recharge. Water levels in shallow open wells were at depths of 2–3 meters below ground in 1986 and by the late 1990s they had dropped to more than 10–15 meters. Prior to the 1990s, seasonal groundwater table, fluctuations were of 2–3 meters, which have amplified to more than

6–8 meters across post-and pre-monsoon. Bailhongal town is also located in the north-eastern portion of the Malaprabha catchment, underlain by hard-rock geology that can retain much less infiltration and thence the recharge, making it more vulnerable to uncertainties for groundwater dependency and forcing greater dependence on the river as the only reliable source of drinking water in the future. Currently, a significant proportion of the water supply system is dependent on the piped water supply system, sourced from the Malaprabha River, throughout the year.

Water Supply Arrangements

Water supply arrangements in Bailhongal town can be broadly categorized as: (*a*) regulated delivery system, (*b*) unregulated private supply system, and (*c*) unregulated open access. Nearly 80 percent of the commercial establishments in Bailhongal benefit from the regulated delivery system, primarily sourced from the river and supplied by the Town Municipal Council of Bailhongal (TMCB). According to the household surveys (conducted during this study), approximately 54 percent of the households are connected to the municipal water supply network; most of these include the residents of the old part of town. While the piped network has extended to new parts of town, supply in the old areas is erratic, partly due to its geographic location (higher elevation). Water tariffs are INR 45 (US$ 1) for residential, INR 90 (US$ 2) for commercial and INR 180 (US$ 4) for construction or industrial uses, respectively. While water prices are different across consumer types, price slabs for additional consumption or cross-subsidized rates across different income categories do not exist, because the supply is unmetered.

Unregulated water markets are also prevalent in the town with many users meeting their water needs by paying private service providers, who supply bulk water in water tankers. Typically, water is delivered at the consumer's doorstep at rates fixed by the private provider (INR 100–200 [US$ 2–4] per tank carrying approximately 2,000–3,000 liters), which fluctuate depending on the level of scarcity and season. Our field investigations revealed that there were at least five private water suppliers in the TMCB area in 2008. Most of them own private bore wells, and distribute water for domestic users in the town. The demand for private water tanker supplies is perennial, with the demand peaking during the pre-monsoon season (February–May). Unregulated free access (a common feature earlier) has now been restricted to those communities typically living near traditional sources of water (mainly open wells or common tanks) or Mini Water Supply Schemes (MWSS) managed by the TMCB.

Water Supply Infrastructure

The major systems of the municipal water supply are the piped water supply (with the river as the primary source), TMCB-owned bore wells and the MWSS. Among the 66 bore wells in the town, 38 serve the MWSS, wherein water from bore wells is stored in tanks and/or supplied, through a piped network, to households or public taps in specific localities or streets. The eight functional open wells (which are privately owned) mostly serve as alternative water supply sources in the old parts of the town, access to which is limited to households residing in the localities. These water supply systems together serve nearly 6,200 households. The piped water supply alone serves 5,200 connections (either individually connected or shared among households). Except for eight of the total 27 wards within the TMCB area, the others are supplied by bore wells and MWSS. Some core areas, mostly inhabited by affluent communities within the old town, receive supplies from both river water and bore wells.

Since its operation in 1976, the piped water supply system in Bailhongal has undergone major infrastructure repairs and a minor upgrade (in 1993) for an estimated population of 25,000 in 2003. Incidentally, since then the operation capacity of the system has drastically declined due to several reasons including population that outgrew the estimates of the original water supply infrastructure design. The river intake system (jack well) does not have a standby pump to ensure uninterrupted operation in case of pump failure. Electricity supply in the last decade has also become highly erratic hampering water uptake, especially during summer when the felt water shortages are most. The capacity of the rising main from the intake point to the Intermediate Pumping Station (IPS) operates at very low efficiency (less than 50 percent) and is insufficient to transmit the raw water, pumped by the 150 HP pump, inducing further inefficiencies in supply operations. Bailhongal is served by a Water Treatment Plant (WTP) with a daily capacity of 0.8 million gallons per day (approximately 3.02 MLD). There is a single elevated reservoir located at the highest elevation point with a 0.2

million–gallon capacity (0.756 million liters), which is less than 20 percent of the total minimum demand of the town at the rate of 60 lpcd for the 47,000 population and 50 percent supply efficiency.

Perceived Water Supply and Shortage

Though the TMCB's Water Supply and Sanitation Department (WSSD) estimates that the current level of piped water supply is around 70 lpcd, the town residents grossly disagree and our household surveys substantiate this discrepancy. We conducted these surveys in Bailhongal town to elicit the nature of water shortage, comprising 310 households across 27 wards, within the TMCB area during 2007–08. The results suggest that during the months of July–December, the water supply was for an average duration of one–two hours, once every week, while the stated demand was at least three hours every week (roughly translating to supply levels of 50–60 lpcd and storage for a week). This worsens during the summer months with the supply lasting for less than an hour, once every two weeks for most households. More than 90 percent of the households experienced at least two hours of supply deficit per week during the summer months (Barton 2008), with some saying that they did not receive any water supply during summer months (see Figure 12.1).

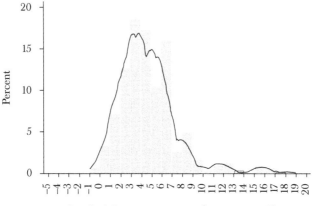

Supply deficit summer season (hours per week)

Figure 12.1: Household water supply deficit relative to stated need during summer

Source: Data from household surveys conducted in the TMCB area.

Our analysis suggests that all 27 wards in the town experienced some form of drinking water shortages round the year, though the magnitude of the problem varied across different wards and communities. More than 14 wards experienced supply deficits of 50 percent or more than the expected level, even during the rainy season, indicating a deficit of 25–30 lpcd. A large number of these households resorted to alternative water supply sources—43 percent use hand pumps, 42 percent are connected to mini water supply (from bore wells), 31 percent use public stand posts, 20 percent use in-house open wells, 5 percent use open private wells, and 5 percent use private bore wells. During drought years, many of these households have no choice but to purchase water from private suppliers.

Irrigation Dimension of Scarcity and Conflict

Similar to most other semi-arid regions in India, agricultural land-use intensification is imminent across Malaprabha sub-basin. With the establishment of several sugarcane factories in the region (and given the growing demand for sugarcane production), the area under cultivation has drastically increased in the past two decades. For example, the net irrigated area in the Belgaum district increased from 10 percent (in 1970) to 30 percent of the total cultivable land in 2001 (according to data from Department of Economics and Statistics, Government of Karnataka) though total cultivated area has fairly remained unchanged. Our own estimates of irrigated area for the year 2006–07 (using time series imagery sourced from IRS-P6 and ground truth points) suggested that the net irrigated area in the Malaprabha reservoir catchment had increased to almost 40 percent of the total cultivable land in the catchment which included sugarcane, paddy, oilseeds, and other irrigated crops (see Map 12.1). Sugarcane, being the most water-intensive and perennially irrigated crop, occupied almost half the irrigated area.

Mapping Water Conflicts on the Ground

In the backdrop of changing cropping patterns, and agriculture-based economic growth, the Malaprabha sub-basin has also witnessed several conflicts over access to water. Cropping pattern violations have consistently occurred in the reservoir command area. Though the reservoir and the canal system was originally designed to provide protective irrigation[1] for an area covering 218,000 hectares, it continues to expand both in the rhetoric of local politics and on ground every year, despite the fact that the reservoir filled just four times in the last 35 years. While not more than 60 percent of the landholding was to be irrigated in any season and the remaining 40 percent was to be either rain-fed or fallow, farmers at the head and mid reaches of the command area continue to grow water-intensive crops and irrigate their entire land holdings. Several unauthorized lands (alongside the canal network) pump water directly from the main canals and distributaries. Issues related to insufficiency and non-reliable supply, at the tail end of the canal command, continue to trigger conflicts in the area. Irrigation management transfer to the local communities, envisaged through Participatory Irrigation Management (PIM) and implemented in the 1990s, yielded no improvement in water sharing mechanism and effective system management (Mukherji et al. 2009).

Farmers in the reservoir catchment (the study area for the chapter) source their irrigation water for annual crops, such as sugarcane, by directly lifting river water (even during summer) over several kilometers. Several Lift Irrigation Schemes (LIS) on the foreshore of the reservoir are either defunct or operate for less than two months due to bad design and insufficient water at the uptake points. Adding to the problem of water shortage and competition, among various users in the sub-basin, are leakages in the supply pipe lines delivering water to Hubli, Dharwad and other towns, such as Bailhongal that primarily depend on the reservoir and the river, respectively, for meeting domestic water demands. The state government of Karnataka is fighting a legal battle with its neighboring riparian state Goa over its proposed inter-basin transfer from Mahadayi River to Malaprabha River, on the grounds that the transfer could help meet the drinking water shortages in these towns, more specifically Hubli–Dharwad. Unlike these twin cities, which have powerful political and other lobby groups for increasing their claims over Malaprabha waters, Bailhongal town is handicapped with its several years of struggle for a stable political system. Severe water scarcity during drought years had forced the TMCB to cart drinking water in tankers from neighboring *talukas* for almost two months. The conflict between Bailhongal drinking water users and the farmers (lifting river water for irrigation) has also reached several flashing points of violence during the last decade. During drought years, the TMCB has repeatedly sought police protection of the river sources in the upper reaches to prevent its pumping from irrigation users.

Events of irrigation pumps being seized and vandalized during conflicts, and enforcement of the IPC Section 144[2] during drought years has worsened the debacle for TMCB, the town residents and inhabitants of neighboring villages who rightfully[3] access the river for their daily domestic water needs and livelihood support.

While the conflict between the irrigation and drinking water sectors, or within irrigation sector across the region, seems inevitable, an unclear rights regime over water resources makes it a further complex and controversial issue for any policy intervention on reallocation of existing water resources. Nevertheless, if the prioritization of the sources of drinking water is implemented, as stated in India's National Water Policy 2002 (GoI 2002b), curtailing profitable uses of water, (for example: irrigating large tracts of sugarcane) becomes justified, especially when such uses are in direct conflict with basic human needs, such as drinking water and subsistence uses (including farming and livestock rearing). Depriving communities of these basic uses of water is a direct violation of fundamental rights if interpreted under the "Right to Livelihood"[4] (also included in the "Right to Life" article of the Constitution of India).

Analyzing Irrigation's Water Capture

In order to assess the physical resource capture by irrigation, an integrated modeling framework, linking socio-economic conditions with bio-physical interactions, was implemented (Reshmidevi and Badiger 2009). The results of modeling scenarios of various land-use change options (see Figure 12.2) suggested that continued diversion of limited water resources for irrigation would continue to reduce river flows downstream, severely affecting availability of water in the river for downstream uses, including drinking water access to towns like Bailhongal. The land-use change scenario analysis in the study addressed two different change patterns (*a*) increasing trends in irrigation from rain-fed cultivation, (*b*) reversing trends of irrigated area conversion to rain-fed cultivation.

Irrigation extraction for sugarcane cultivation drastically reduced the stream flow, especially during the dry season. Simulation results indicated that the intensity of physical water scarcity increased by an additional 35 percent, when 56 percent of the current rain-fed areas were converted to sugarcane. This degree of scarcity was lesser at lower rates of land conversion to sugarcane. In the second set of simulations, stream flow predictions were made for land-

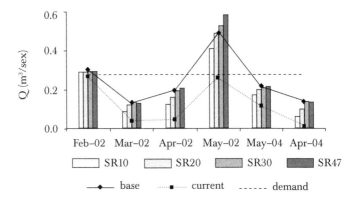

Figure 12.2: Impact of land-use conversion from irrigated to rain-fed crops

Source: Based on outputs of various scenario simulation.

use change from sugarcane land-use to rain-fed cropping systems. Mathematically, this shift is equivalent to maintaining the same area under sugarcane with improvements in irrigation efficiency measures by the same percentage. For the years 2002 and 2004, where the annual rainfall was below the long-term annual average rainfall (similar to meteorological drought conditions), significant increase in the dry season stream flow was observed when 10 percent of the existing sugarcane areas were converted to rain-fed cultivation or when irrigation efficiency improvements were increased by that order. Under this scenario, the duration of water scarcity was found to reduce from four months to two months. In addition, an average 50 percent reduction in the intensity of scarcity was estimated when 47 percent of the sugarcane areas were converted to rain-fed crops.

Rethinking Urban Water Governance

The TMCB-WSSD is headed by a Junior Engineer (JE) with a support staff who maintains and operates the town's water supply infrastructure. A comparison of the current level of staffing and the required staffing numbers as expressed by the WSSD (see Table 12.1) reflects the fact that the department is highly understaffed to carry out the daily duties of system operation and maintenance.

Table 12.1: Current and recommended staffing for the
Water Supply Department

Staff Position	Recommended staff (no. of persons)	Present staff (no. of persons)
Junior Engineer	2	1
Maintenance staff	16	11
Senior operators	2	0
Operators	13	0
Assistant operators	13	4

Source: Data collected from personal interviews.

Under the provisions of the Karnataka Municipalities Act (1964), the Bangalore Water Supply and Sewerage Board Act (1964) and the Karnataka Municipal Corporations Act (1976), the ULBs, such as the TMCB, are obligated to supply potable, wholesome water to the citizens residing in their respective jurisdiction. In Bailhongal, the TMCB's WSSD manages the water supply, along with the elected body of councillors and the TMCB's Revenue Department. Decisions regarding planning, design and construction of all major works, estimated to cost more than INR 1.5 million (US$ 25,000), are routed through the Karnataka Urban Water Supply and Drainage Board (KUWSDB), while works related to irrigation structures, are decided by the Karnataka Neeravari Nigam Limited (KNNL). The WSSD is also responsible for all decisions regarding operation and maintenance (O&M) of the water supply system including piped water supply from the Malaprabha River and bore well supplies through MWSS. Works costing up to INR 20,000 (US$ 445) require the approval of the TMCB's Standing Committee on Public Works, while those with higher cost estimates require the approval of the General Body of councillors. The irregularity of convening the Standing Committee or the General Council for work approval often prolongs responding to system complaints. To avoid such delays in sanctioning, the most urgent work is sub-divided into parts of INR 5,000 (US$ 110) and is implemented by the water supply department without consultations with the Council or the Standing Committee. Even though this allows for quicker response to system complaints, very often it does not provide a sustainable solution to plug the actual system inadequacy and inefficiency. For example, most complaints of poor water availability are solved by ordering more bore wells to be dug in an already depleting aquifer. Due to poor institutional integration between various departments for effective O&M, which by law is the mandate of the ULB, proper participatory planning goes unaddressed during implementing new works. This often leaves the ULB with assets that are difficult to manage, which further bring down the financial stress and staff capacities of the ULB to perform functions, even outside the routine water supply services.

Efficiency and Equity

As noted earlier, the WSSD charges a flat rate for water consumption across all economic categories of households but this does not reflect the utility value for water across these. Due to far lower water storage capacities and low ability to pay for municipal water supply, water has higher utility value among the lower-income groups (LIGs). However, these communities are more vulnerable and derive much less benefits when compared to the economically well-off who, by building high storage capacities, could use more water than necessary and are far less vulnerable to erratic water supply. In certain areas (wards 1, 2, 9 and 26) that have a majority of consumers from the low-income group, the preference for service provider is shifting to private tanker supplies. With most tankers supplying water at INR 100–200 (US$ 2–4) per tank and the average demand assumed to be at 20 lpcd (less than the calculated average), these consumers end up paying nearly three times more a month to meet their water demands.

Our analysis of the coverage of the different systems of water supply in the town suggests that nine wards out of 27 have poor coverage (wards 6, 8, 15, 22, and 26 in the inner zone; wards 9, 10, 13, and 23 in the outer zone). Pipe connections per hundred households varied from 22 to 100 across the 23 wards that existed in 2004. With increasing water demands, private suppliers would prefer consumers with better ability to pay, thereby increasing the existing distributional inequities. Absence of differential pricing across economic categories of users makes the state delivery system less adaptable to changing economic conditions, such as increased operating costs due to market fluctuations or increased capital expenditure due to increased infrastructure demand, or to changing resource conditions like rapid growing consumer numbers, drought conditions, among others.

The piped water supply has the highest coverage among the various arrangements for water supply in Bailhongal. From the analysis of the WTP survey data, we found that the average quantity of water delivered is between 15–20 lpcd, whereas the production of water at the intermediate pumping station ranges between 300 lpcd, in the dry season, to 330 lpcd, during normal months of the year. This leaves huge scope for lessening the degree of water shortage in Bailhongal through system improvements. The benefits of investing in improvements in the internal system performance could be better distributed among the consumers rather than investing the same amounts on transmitting water from a source as far as seven kilometers from the town and introducing the same into an inefficient distribution system.

Accountability

In the state delivery system, the elected body of councillors represents the consumers in decision-making. However, from interviews conducted among consumers and nine of the 27 ward councillors, it is understood that the functioning of the body is driven by political affiliations. Processes like the appointment of the chairperson of the Standing Committee are erratic and non-democratic, entirely based on a nomination by the president of the TMCB. The rent-seeking behaviour of councillors has also forced consumers to depend on the WSSD officers with their grievances. For the same reasons, maintenance works are divided into estimates below INR 5,000 (US$ 110), so that their implementation can be carried out with minimum intervention from the general body of councillors. Such exclusion of the TMCB, however, adds to the lack of accountability (such as critical decisions regarding fund allocations and choice of maintenance works) of the WSSD. Since the WSSD or the councillors do not directly reap the benefits of the service, and external funding is available only on new works, the incentive for the ULB to invest in maintenance or ensure financial recovery for sustainable operation and maintenance works is also poor. The power structure is heavily inclined in favour of representatives with strong political affiliations (historically or otherwise). The flow of information among different ward representatives and consumers is skewed and, sometimes, distorted. Women representatives, who were interviewed, expressed lack of awareness of water supply proposals and poor satisfaction on the council's responses to their demands to ensure better water supply in their respective wards. Even male councillors (though some were aware of the new proposals) had little information regarding the present status of work on these. Neither the politically elected representatives nor the WSSD (supposedly an apolitical and independent implementing body) seem to represent the interests of the various water users in Bailhongal.

From our analysis, it is evident that the municipal council of Bailhongal town and its governing bodies seem abysmally unprepared to address the problem of water scarcity in the town. Though the needs of the town are basic, they have been complicated due to the intertwined nature of the socio-political system. Reform elements of the governing structures require systems for (*a*) financing that allow the local urban body to undertake basic tasks of ensuring water supply; (*b*) providing the space for citizens and elected representatives to participate in effective governance; and (*c*) sorting out systemic problems so that local councils can move ahead with developmental plans. Building the capacity of the urban local bodies and the elected representatives seems critical so that they use their capacity and powers more efficiently and effectively.

Conclusion

Domestic water scarcity is as much a socio-institutionally driven outcome as growing physical scarcity of water due to competition over limited water resources, and is largely constructed at regional and local scales. Solutions to such problems, hence, require understanding the inter-connectedness of bio-physical and socio-economic issues at the micro, meso and macro scales, and require integrated planning to avoid conflicts. At the regional scales, policy decisions for agricultural land-use regulation are critical to ensure adequate water allocation for domestic needs from surface and groundwater resources. At local scales, such as small urbanizing towns, rebuilding the trust of water users in the ULBs seems essential. Planning of water supply service improvements should stress on the distributional issues of the water supply infrastructure and effective institutional functioning that largely depend on the co-ordination between the politically elected and executive functionaries in the governance structure, rather than building additional capacities to pump or divert more water from alternative sources. An institutional and legislative platform for enabling negotiating environments between competing and conflicting water uses becomes essential, particularly when water rights are unclear. Institutional reforms to improve equitable distribution, accountability of the service providers, and adaptability to changes in resource and economic conditions hold great promise for arriving at environmentally and socially sustainable solutions.

Notes

1. Protective irrigation is a water allocation system that spreads scarce water resources over a considerable area and large number of farmers to protect them against total crop failure. Though formulated during the Colonial period, this system has dominated major irrigation policies in the post-colonial, planned economic development period. The primary goal is to distribute water to as large a number of farmers as possible, thereby minimizing the risk in cultivation. Economically, it may or may not be the most optimal distribution for achieving highest gross water efficiency or productivity. However, in reality, farmers seeking to maximize the productivity of land often violate cropping patterns recommended in accordance with the principle of protective irrigation. See Jurriens et al. (1996) for other interpretations of the terminology in different contexts.

2. Section 144 of the Indian Penal Code (1860) states:
Joining unlawful assembly armed with deadly weapon refers to whoever, being armed with any deadly weapon, or with anything which, used as a weapon of offence, is likely to cause death, is a member of an unlawful assembly, shall be punished with imprisonment of either description for a term which may extend to two years, or with fine, or with both.

3. Under the Riparian Water Rights principle, all landowners whose property is adjoining a body of water, such as a river, have the right to make reasonable use of it (Caponera 1992). Also, according to India's National Water Policy (GoI 2002b), drinking water needs are prioritized over other uses including agriculture and industry.

4. Right to Life and Right to Livelihood: Article 21 of the Constitution of India (1950: 14–20) provides, "No person shall be deprived of his life or personal liberty except according to procedure established by law." "Life" in Article 21 of the Constitution is not merely the physical act of breathing. It does not connote mere animal existence or continued drudgery through life. It has a much wider meaning which includes right to live with human dignity, right to livelihood, right to health, right to pollution-free air, among others. Right to life is fundamental

to our very existence without which we cannot live as human beings and includes all those aspects of life, which together make a man's life meaningful, complete and worth living.

References

Barton, D. 2008. "Evaluating Feasibility of Pricing Domestic Water Supply Using the Contingent Valuation Method." CISED-NIVA Technical Brief No. 6–2008. Available at http://www.malaprabha.org/publications/TechBrief6.pdf (accessed December 8, 2008).

Caponera, D. A. 1992. *Principles of Water Law and Administration: National and International*, Vol. 1. Rotterdan: Balkema.

Constitution of India, 1950. *Article 21*. Delhi: Metropolitan Book Company Ltd; Law Booksellers and Publishers.

Falkenmark, M. and C. Widstrand. 1992. *Population and Water Resources: A Delicate Balance*. Population Bulletin. Washington, DC: Population Reference Bureau.

Government of India. 1860. *The Indian Penal Code*, Section 144. Available at http://books.google.co.in/books?id=-MsSAAAAYAAJ (accessed March 1, 2010).

———. 2002a. "India Assessment 2002: Water supply and Sanitation," A WHO–UNICEF-sponsored study, Planning Commission, GoI. Available at planningcommission.nic.in/reports/genrep/wtrsani.pdf (accessed March 1, 2010).

———. 2002b. *National Water Policy*. Available at http://mowr.gov.in/writereaddata/linkimages/nwp20025617515534.pdf (accessed March 1, 2010).

———. 2006. "Report of the Working Group on Water Resources for the XI the Five Year Plan (2007–12)," Ministry of Water Resources. Available at http://planningcommission.nic.in/aboutus/committee/wrkgrp11/wg11_wr.pdf (accessed March 1, 2010).

Government of Karnataka. 1964. *Karnataka Municipalities Act*. Available at http://www.lawsofindia.org/pdf/karnataka/1964/1964KR22.pdf (accessed March 1, 2010).

———. 1964. *Bangalore Water Supply and Sewerage Board Act*. Available at http://www.lawsofindia.org/pdf/karnataka/1964/1964KR36.pdf (accessed March 1, 2010).

———. 1976. *Karnataka Municipal Corporations Act*. Available at http://dpal.kar.nic.in/pdf_files/14%20of%201977%20(E).pdf (accessed March 1, 2010).

Jurriens, M., P. Mollinga, P and P. Wester. 1996. *Scarcity by Design: Protective Irrigation in India and Pakistan*. Report by Wageningen Agricultural University and International Institute for Land Reclamation and Improvement the Netherlands. Available at http://www2.alterra.wur.

nl/Internet/webdocs/ilri-publicaties/special_reports/Srep4/Srep4.pdf (accessed March 1, 2010).

Kundu, A., S. Bagchi and D. Kundu. 1999. "Regional Distribution of Infrastructure and Basic Amenities in Urban India: Issues Concerning Empowerment of Local Bodies," *Economic and Political Weekly*, 34(28): 1893–1906.

Mehta, L. 2003. "Contexts and Constructions of Water Scarcity," *Economic and Political Weekly*, 34(48): 5066–72.

———. 2007. "Whose Scarcity? Whose Property? The Case of Water in Western India," *Land Use Policy*, 24(4): 654–63.

Mukherji, A., T. Facon, J. Burke, C. de Fraiture, J. M. Faures, B. Fuleki, M. Giordano, D. Molden, and T. Shah. 2009. *Revitalizing Asia's Irrigation: To Sustainably Meet Tomorrow's Food Needs.* Colombo: International Water Management Institute (IWMI) and Food and Agriculture Organization (FAO).

Ohlsson, L. and A. R. Turton. 2000. "The Turning of a Screw: Social Resource Scarcity as a Bottle-Neck in Adaptation to Water Scarcity," in *Stockholm Water Front–Forum for Global Water Issues*, No. 1 (February). Stockholm: Stockholm International Water Institute (SIWI).

Reshmidevi, T and S. Badiger. 2009. "Impact of Irrigation Intensification on Inter-Sectoral Water Allocation in a Deficit Catchment in India," in Günther Blöschl, Nick van de Giesen, D. Muralidharan, Liliang Ren, Frédérique Seyler, Uttam Sharma, and Jaroslav Vrba (eds), *Improving Integrated Surface and Groundwater Resources Management in a Vulnerable and Changing World*, pp. 61–68. International Association of Hydrogeologists (IAHS) Publication No. 330.

Shiklomanov, I. A. 1998. *World Water Resources: An Appraisal for the 21st Century. International Hydrological Programme* (IHP) Report. Paris: UNESCO.

13

Urban–Rural Water Nexus

The Case of Gujarat

R. Parthasarathy and Soumini Raja

The state of Gujarat is divided into six agro-climatic zones, namely North Gujarat, South Gujarat, Central Gujarat, Saurashtra, and Kachch. A large variation in the availability of water across the state has a direct correlation with the geography, rainfall pattern, changing land use, and management practices. In terms of the total quantum of rainfall, South and Central Gujarat receive the highest. Gujarat's total water resource potential is 50,000 million cubic meters (mcm), of which surface water is about 38,000 mcm (76 percent) and groundwater is about 12,000 mcm (24 percent).

The major part of surface water is available through the 185 river basins. The water is used mainly for irrigation, drinking as well as industrial purposes. Of these 185 river basins, major basins of Sabarmati, Mahi, Narmada, and Tapi drain the central and southern parts of the state while the minor basins (grouped as westerly flowing rivers group 1) are concentrated in the Kachchh and Saurashtra regions (see Map 13.1). Over the years Gujarat has become one of the most rapidly growing states in industrial investment and economic development causing stress on the resources, especially water.

In the state, there are about 150 urban centers,[1] of which seven are under municipal corporations and the rest are under municipalities.

River Basin Boundaries
State Boundaries

River Basins

1. Indus
2. Mahi
3. Narmada
4. Sabarmati
5. Tapi
6. Westerly Flowing Rivers–
   Group 1
7. Westerly Flowing Rivers–
   Group 2
8. Brahmani and Baitarani
9. Cauvery
10. Easterly Flowing Rivers–Group 1
11. Easterly Flowing Rivers–Group 2
12. Ganga
13. Godavari
14. Krishna
15. Mahanadi
16. Pennar
17. Subarnarekha
18. Brahmaputra
19. Meghna

Map 13.1: River basins in India

Source: Amarasinghe et al. 2004.

Approximately, 38 percent of the population lives in the urban areas. An analysis of the literature on river basin studies in Gujarat (Parthasarathy 2009) shows that the key challenge faced by the state today is to bring about a balance between the water demands and consumption in urban and rural areas. There has been a constant deterioration of surface water, especially in the Sabarmati and Tapi river basins, owing to the large scale investment in industries and rapid urbanization trends.

There has also been a visible decline in both surface water and groundwater quality in various parts of the state. The major reasons attributed to this phenomenon are the increasing density and deepening of tube wells, dumping of waste water by the local authorities and discharge of untreated or partially treated effluents by the industries into the rivers and other water bodies. The agricultural run-off and disposal of hazardous waste (in different forms) in landfills near the urban areas and industrial zones have also affected the quality of surface and groundwater.

The polluted water reduces the net utilizable safe water and aggravates the problem of water scarcity largely in the rural areas of Gujarat, especially those in close proximity to the urban and industrial centers. This chapter investigates this urban–rural water nexus in the Sabarmati and Tapi basins–which account for some of the most important urban and industrial centers in the state.

Urban Centers in the River Basins

The Golden Corridor,[2] a major driver of economic development, passes through the central and lower reaches of Sabarmati and Tapi river basins. The basins comprise some of the most important urban centers, industrial clusters and productive agricultural lands that contribute to the State's Domestic Product. However, the careless and insensitive management of these centers has resulted in the large-scale deterioration of the waters in these river basins.

Domestic withdrawals consist of mainly two components–water withdrawals for human consumption (including domestic services), and water withdrawals for livestock. According to the Central Water Commission (CWC) estimates in 2002, the human demand for drinking, cooking, bathing, recreation, and other purposes accounts for 79 percent of domestic withdrawals (Amarasinghe et al. 2004). Water demand in urban areas is higher due to water use for flushing latrines, gardening and fire-fighting, among other uses.

An increasing consumption of water by the domestic and industrial sectors over the years has subsequently led to the increasing release of untreated waste water into the natural water courses. The most noticeable implication of this process is the severe damage caused to natural resources and to a large extent, the agricultural areas. Though waste water acts as a substitute for chemicals in agriculture, it has often proved fatal not only to the productivity of crops but also to the health of the producers (Drechsel et al. 2009). More importantly, with the increasing consumption of water (surface and ground sources) by urban centers, the demands of the rural areas are often not met, increasing their dependence on polluted sources.

River basin based studies may help in bringing out this overarching issue of urban–rural water nexus, especially in the context of Gujarat. With more than 54 to 63 percent of the population classified as rural, in the Sabarmati and Tapi basins respectively, it would be important to take a closer look through an assessment of the prevailing conditions in the major urban centers such as Ahmedabad (Sabarmati River basin) and Surat (Tapi River basin) and their rural peripheries.

Case 1: Sabarmati River Basin

Sabarmati River Basin is categorized as a water-deficit basin which lies on the west coast of India, between latitudes 22° N to 25° N and longitudes 71° E to 73° 30' E, and is spread across the States of Rajasthan and Gujarat. Sabarmati River originates at an altitude of 782 m in the Aravalli Hills in Udaipur (in Rajasthan) and flows for a length of 371 km in the south-west direction, of which 48 km are in Rajasthan and 323 km in Gujarat. The river joins the Gulf of Khambat in the Arabian Sea (Gopalakrishnan et al. n.d.).

The major characteristics of the Sabarmati and Tapi basins are given in Table 13.1. The Sabarmati basin has a total drainage area of 21,565 Sq km, of which 17,441sq. km is in Gujarat and 4,124 sq. km is in Rajasthan (Gopalakrishnan et al. n.d.). Approximately 57 percent of the total land use is under agriculture while about 20 percent is not under cultivation. The Sabarmati basin covers parts of the districts of Banaskantha, Sabarkantha, Mehsana, Gandhinagar, Ahmedabad, and Kheda. There are about 20 industrial estates located within the river basin and of the many urban centers, the cities of Ahmedabad and Gandhinagar are located along its central reaches.

Ahmedabad

Ahmedabad is located in the south-western part of the Sabarmati basin, occupying 27.3 percent of the total area. Sabarmati River

Table 13.1: Characteristics of Sabarmati and Tapi river basins

Attributes	Sabarmati	Tapi
1. Catchment Area (sq. km)	22	65
2. Length of the river (km)	371	724
3. Total Population (millions)	6	17.9
4. Density of Population (No./sq. km)	521	245
5. % Rural Population	54	63
6. Total Renewable water resource (cu. km)	3.8	14.9
7. Potentially utilizable water resources (Surface water) (cu. km)	1.9	14.5
8. Potentially utilizable water resources (Ground water) (cu. km)	2.9	6.7
9. Potentially utilizable water resources (Total water) (cu. km)	4.8	21.2
10. Total Renewable water resource available per capita (cu. m)	631	831
11. Potentially utilizable water resources available per capita (cu. m)	797	1183

Source: Amarasinghe et al. 2004.

comprises three sub-basins–Dharoi, Watrak and Hathmati (Winrock International India 2006). The Ahmedabad Urban Agglomeration spreads over 1330.08 sq. km; 190.84 sq. km area falls within the jurisdiction of the Ahmedabad Municipal Corporation (AMC) and 150 villages under the jurisdiction of the Ahmedabad Urban Development Authority (AUDA). The Sabarmati River divides the city into the old city and its periphery on the eastern bank and the new city on the western bank.

In 2001, the AMC accounted for a population of 35,20,085 with a density of 184 persons per hectare and the AUDA accounted for 47,09,180 with a density of 77 persons per hectare. According to the 2001 census, the city accounts for seven percent of the state's population and 20 percent of its urban population (CEPT 2006a). Also, Ahmedabad city accounts for 21.5 percent of factories in the state, employing 18 percent of workers. Currently, there are around 4,859 factories in the city of which chemical and petro-chemical industries have the largest share of 29 percent and textile industries, 13 percent and these are considered to fall within the highly polluting category. However, approximately 20 percent of the land use in the AUDA (including the AMC) is classified under agricultural use while industries account for about 17 percent (14.3 in the AMC and 3 in the AUDA, excluding AMC) (CEPT 2006a).

The water supply needs of the urban area are met from three sources: (*a*) surface water from Raska pipeline, (*b*) French well in Sabarmati River and (*c*) by the intake well, constructed in the river. The Narmada canal, which passes through the north of the city, releases water that is pumped through the intake wells. The AMC also draws water from 363 bore wells installed in various parts of the city.

Earlier studies have shown that though the average daily supply of water per capita has increased from 20.24 million gallons, in 1951, to 104.83 million gallons, in 2001, the gross per capita per day has reduced over the years (see Table 13.2) (WinrockIndia 2006). According to the AMC, in 2005, of the total 590 mld supplied, 492 is for domestic purposes, 66 for commercial and industrial purposes and 30 is delivered to the public stand posts. However, in the peripheral areas outside the jurisdiction of the AMC, organized water supply is limited largely to *gamtal* (village settlement) areas that cover less than 10 percent of the population. With increasing urbanization pressures and demand on infrastructure within the city limits, the peripheral areas are not catered to and, subsequently, are forced to depend on bore wells. These villages face problems that are dual in nature–one, they are associated with excessive withdrawal of groundwater and deterioration of water quality due to fluoride intrusion and, two, the use of surface water, which is often contaminated with pollutants from industries, is affecting agriculture and livestock productivity (see Table 13.3).

Table 13.2: Water supply trends in Ahmedabad (1951–2001)

S. No	Year	Population	Average daily supply (Million Gallons)	Gross Per Capita per day Gallons (per day)
1.	1971	15,85,544	69.98	44.14
2.	1981	20,59,725	96.08	46.64
3.	1991	28,76,710	93.27	32.42
4.	2001	35,15,361	104.83	29.82

Source: Winrock International 2006.

Table 13.3: Water quality test result in Sabarmati Basin

Village/Source	Cadmium	Chromium	Copper	Lead	Zinc	Arsenic
Permissible Limits for heavy metals in water (ISI)	0.01	0.05	1.5	0.10	15.00	0.00
Galiyana River Water	0.02	0.46	0.16	0.00	0.07	0.00
Sahij River Water	0.02	0.49	0.28	0.16	0.08	0.00
Gyaspur River Water	0.007	0.92	1.57	0.17	0.65	0.00
Vautha Borewell	0.01	0.61	0.22	0.40	0.56	0.00
Sahij Soil	0.15	25.04	19.4	0.98	36.59	0.00
Vautha Wheat	0.00	0.00	0.00	2.675	0.00	0.00

Source: Winrock International India 2006.

Currently, there are 20 industrial estates in the Sabarmati basin, out of which three–Odhav, Naroda and Vatva–are located within the Ahmedabad district boundary. Vatva has 1,750 industries, Naroda, 850 and Odhav, 765 (WinrockIndia 2006). Three Common Effluent Treatment Plants (CETPs) have been installed in the industrial estates and the treated sewage is discharged into the river. The less and non-polluting industries, under the Gujarat Industrial Development Corporation (GIDC), are not treating the waste (at present) and the effluent is discharged either in open or in the Khari canal. The other industries in the AMC and Behrampura (which are highly toxic) are connected to the main sewer line and, here too, the effluents are being discharged into the river without any treatment (CEPT 2006a). There are 27 storm water drain outlets into the river within the Ahmedabad Urban Agglomeration. Within the urban area, domestic sewage of 500 mld is generated, of which about 168 mld is discharged into the river through storm water outlets, without treatment (CEPT 2006a). An analysis of the land use within the river basin shows that in approximately 13 percent of the area, double cropping is practiced and this patch of agricultural land lies within the lower reaches of the basin (CEPT 2009).

The annual mean water resource in the basin is estimated as 3,810 million cu. m. The total demand for the year 2001 was 5,744 million cu. m. The consumption of surface water for irrigation is estimated to be 3,465 million cu. m per year, including the Mahi command within the Sabarmati basin (1,663 million cu. m), while the groundwater contribution to agricultural use is estimated as 2,279 million cu. m. Water requirement for humans and livestock (in 2001) was 510 million cu. m (Gopalakrishnan et. al. n.d.) (see Table 13.3).

A sample study of villages lying in the south of Ahmedabad Urban Agglomeration, such as Vautha, Sahij and Navagam Karna, by CEPT University and of Asamli, Bakrol, Chitrasar and a few others by Winrock International India, has pointed out that these villages use this polluted river water for irrigation and for livestock. The quality assessment of water samples has shown a large concentration of lead, cadmium, chromium, and other metals; their concentrations were well above permissible limits (see Figure 13.4) (CEPT 2009; Winrock International India 2006).

Recent studies (Winrock International India 2006; CEPT 2009) revealed that the impact could be classified based on agricultural productivity and practices, human and livestock health, and impact on fauna and flora. A common observation in both studies has been the occurrence of high TDS (total dissolved solids) in the bore wells. The Central Pollution Control Board (CPCB) in 2007 rated Sabarmati river water in category "E"[3] with some of the

Table 13.4: Irrigation withdrawal in river basins

Sl No	Attributes	Sabarmati	Tapi
1.	Withdrawal per person (cu. m)	573	381
2.	Net irrigated Area (Mha)	0.36	0.64
3.	Irrigation Intensity (%)	122	120
4.	Groundwater irrigated area (%NIA)	90	64
5.	Grain Crop Irrigated area (% NIA)	38	47
6.	Overall irrigation efficiency	60	55
7.	Potential annual evapo–transpiration (mm)	1947	1890
8.	Annual 75% dependable rainfall (mm)	384	455
9.	Crop water requirements (mm)	455	452

Source: Amarasinghe et al. 2004.

Table 13.5: Labor and irrigation in the Surat district *talukas*
located in the lower Tapi basin

Talukas (R)	% rural population	% cultivators	% agri. laborers	% Land under cultivation	% Land Irrigated
Olpad	93	17	39	56	53
Kamrej	100	10	55	65	73
Surat City	0	0	0	0	0
Chorasi	34	7	20	27	28
Palsana	77	8	58	66	72

Source: Census of India 2001.

downstream stretch (such as Ahmedabad to Vautha) as below "E."
It was observed that many of the villagers depend on the waste
water for irrigation purposes which, to a large extent, has resulted
in chemical contamination of fruits and grains and has also led to
lower yield of certain crops such as wheat, jowar and chana. Various
health implications, such as intestinal problems, skin irritations and
joint pains, have mainly affected agricultural laborers and lower
caste members of villagers, who are in direct contact with the river
water. Various livestock disorders have also been reported due to
drinking polluted water. Animals showing weakness at a very early
age, frothing at the mouth, swelling of throat, and foot rot have
been reported from these villages. Moreover, milk production has
dropped drastically (Rural Component Lab Report 2009).

Case 2: Tapi River Basin

The south Gujarat region is drained by three, major west-flowing
perennial rivers namely, Mahi, Narmada and Tapi. The region
covers 24 percent of the state's geographical area and receives 40
percent of the precipitation.

The Tapi (see Figure 13.4) is one of the major rivers flowing
westward through the states of Madhya Pradesh (9,804 sq. km),
Maharashtra (51,504 sq. km) and Gujarat (3,837 sq. km), draining
into the Gulf of Khambat in the Arabian Sea. The Tapi river basin
extends up to 65,145 sq. km and is the ninth largest in the country.
The basin can be classified into two parts: (*a*) the upper region that is
predominantly agricultural land, falling within the state of Madhya

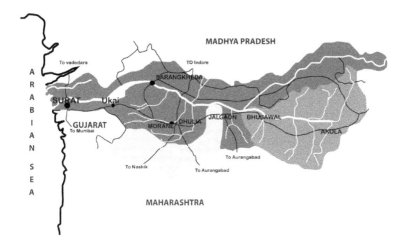

Map 13.2: Tapi river basin

Source: Ministry of Water Resources.

Pradesh, and (*b*) the central and lower reaches, flanked by heavy industries in the states of Maharashtra and Gujarat. Some of the major urban centers, such as Burhanpur in Madhya Pradesh; Akola, Bhusawal, Jalgaon, Malegaon, and Dhule in Maharashtra; and Surat in Gujarat–are located within the Tapi basin area.

Surat Urban Center

Unlike the Urban Agglomeration of Ahmedabad, located in the south-western reaches of the Sabarmati basin, the Surat Urban Agglomeration is located along the mouth of the Tapi River. However, increasing urbanization trends, migration and heavy industries have put an immense pressure on the neighboring villages, located on the immediate periphery of the urban center, especially those located along the coast. The Surat Urban Agglomeration, under the Surat Urban Development Authority (SUDA), consists of Surat city and 148 villages. The city is governed by the Surat Municipal Corporation (SMC) covering an area of 326.52 sq. km, extending up to the Arabian Sea coast, and the SUDA that covers an area of 722 sq. km (see Figure 13.5). The Surat Urban Agglomeration extends over Surat city, Olpad, Chorasi, Kamrej, and Palsana *talukas*, as well as the Hazira Industrial Area (168 sq. km).

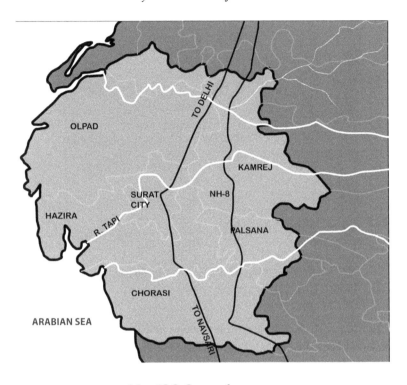

Map 13.3: Surat urban center

Source: CEPT 2006b.

The city of Surat is India's 12th and Gujarat's second most populous city. In 2001, Surat (within the jurisdiction of the SMC) had a population of 24,33,785 with a density of 217 persons per hectare–before expansion of city limits from 112.27 sq. km (in 2001) to 326.52 sq. km with a density of 88 persons per hectare (in 2006)– and SUDA with a total population of 30,90,686. It is known for its industrial base, which plays an important role in the economy of the state, and for its textile manufacturing, trade, diamond cutting and polishing industries, intricate *zari* works, chemical industries, and gas-based industries at Hazira, established by ONGC, Reliance, ESSAR, and Shell. There are more than half-a-million industrial units, mainly consisting of textile industries. Textile units depend on groundwater for textile processing and withdraw about 700 to

1000 cu. m of water every day. There are six main industrial estates located within the city in Pandesara, Khatodara, Udhana, Katargam, Sachin, and Bhestan, which have more than 30,000 industrial units (CEPT 2006b).

The urban center has seen a consistent growth in its population, which has been mainly attributed to the employment opportunities in the fields of manufacturing and other sectors (ibid.). Consequently, the increasing demand for residential, commercial and industrial use has put enormous pressure on its natural resources especially the water supply. The main source of water for the city has been the Tapi River for centuries. The contribution of groundwater is only about 10 percent of the daily total water supply to the city. At present, the SMC supplies water to 95 percent of its population. The total water supplied, including ground and surface sources, within the city is 580 mld (supply capacity is 6,732 mld). 502 mld is supplied for domestic purposes, 20 for commercial purposes and 55 for industrial purposes.

However, the area that is not served is largely made up of agricultural land within the SMC. Rural areas within the urban periphery face a similar problem. Most of the villages located along the periphery depend for their livelihoods on agriculture and fishing (Parthasarathy and Soumini 2009). The SUDA area (except the SMC) has no water supply system of its own. The main source of water for various villages is groundwater that is tapped through bore wells or the water supply schemes supported by Gujarat Water Supply and Sewerage Board.

According to the Gujarat Pollution Control Board, the quality of the Tapi River has been classified based on parameters like BOD, DO and pH data, and shows signs of pollution. The deterioration of the surface water quality also affects the groundwater quality, which can be further attributed to the dumping of waste water by local authorities and the discharge of untreated or partially-treated effluents (by the industries) into the rivers and other water bodies. The agricultural run-off and disposal of hazardous waste in different forms in landfills, near the urban areas and industrial zones, also affect the quality of surface and groundwater. The polluted water reduces the net utilizable safe water and aggravates the problem of water scarcity. Unregulated lifting of groundwater (through bore wells) coupled with close proximity to the coast has increased the salinity of groundwater. The Surat district is affected partially by salinity, fluoride and nitrate contents in ground water. The percentage of villages affected by salinity is highest in south Gujarat (6.5 percent), followed by middle Gujarat (6 percent), which is mainly attributed to the proximity to the sea (Parthasarathy and Soumini 2009). A survey in the coastal villages of the district has shown that irrigation plays a major role in the agriculture of this region (ibid.). In this district, 60 percent of the total irrigated area is under canal irrigation, which channels the water from local perennial streams and rivers to agriculture fields. The district has 54 percent of irrigated land to net sown area as compared to 32 percent for the state as a whole.

Industrial and urban development in the lower Tapi river basin has posed various challenges not only to the livelihood (agriculture and fishing) in the neighboring villages but also to the fauna at the mouth of the river. Site visits to the coastal villages of Chorasi and Olpad *talukas* have revealed that with the increasing threats from future changes in climate, these unplanned developments aggravate the vulnerability of coastal rural livelihoods (ibid.).

A field study was conducted in 2009, taking into account 75 households spread across 10 coastal villages. Villagers complained about the impact of the existing industrial cluster at Hazira on the annual productivity of crops. It has been observed that there is a reduction in the size of fruits and vegetables over the past decade, mainly attributed to the increasing chemical content in soil and air and other direct pollution from the industries.There is a visible increase in salinity in many fields, especially in thevillages of Olpad *taluka*, affecting the availabilty of water for irrigation. There has also been an expansion of city boundaries, over a period of five years, which has led to the conversion of large tracts of agricultural land to non-agricultural use. Many agricultural workers are shifting to non-agricultural work such as industries or factories in the nearby areas. Most of these villages are located on the tail-end of canal irrigation systems; this has, in turn, further affected the availability of water for irrigation. A large number of farmers depends on wells for irrigation purposes.

The location of industries along the mouth of the Tapi River has affected the livelihood of fishermen drastically–who've reported a tremedous decline in the catch of varieties of fish such as Pomphret and Chilia,when compared to the past decades. The industries located along the coast and rivers have been reported as polluting the water bodies harming fish breeding and rearing. In 2008, there were repeated dead fish incidences along Dumas Beach. Moreover, conversion of mangrove patches and denotifying them for industrial and port activities have also affected the fish breeding in the area. Large scale dredging in the river mouth has also affected the aquatic ecosystem (Parthasarathy and Soumini 2009).

Conclusion

The river basin studies have helped in bringing out the overarching issue of the urban–rural water nexus especially in Gujarat, which is currently the most industrialized state in India. With more than 54 and 63 percent of the population classified as rural in the Sabarmati and Tapi basins, respectively, this chapter attempted an assessment of the prevailing conditions in the major urban centers such as Ahmedabad (in the Sabarmati river basin) and Surat (in the Tapi river basin) and their rural peripheries.

The case of the Sabarmati River in Ahmedabad has revealed that the discharge of treated and untreated waste and toxic effluents into the river has impacted various villages located in the downstream areas. In addition, the domestic sewage of about 168 mld from the city is also discharged into the river without treatment. With approximately 13 percent of the river basin area under double-cropping, mainly in the lower reaches, the impacts of industrial and urban domestic waste have turned critical in the rural areas. They have been affected not only by decreasing productivity but also by the contamination of the crops and land, thereby impacting the health of humans and livestock. However, the case of the Tapi river basin revealed a different pattern of the urban–rural nexus. Although both showed similar trends in the deterioration of the surface water quality, the groundwater was influenced by the release of waste water by urban authorities and the discharge of untreated or partially-treated effluents by the industries into the rivers. In Surat (due to its proximity to the Arabian Sea), water resources have been affected by both anthropogenic activities in urban and industrial centers as well as increasing salinity. The impact of increasing industrial and urban density has not only affected the agricultural sector but other primary sectors, such as livestock rearing and fishing as well. These two cases have revealed a strong evidence of an urban–rural nexus which often remains unidentified and is overlooked due to the existing planning processes that are piecemeal in their nature and approach. This emphasizes the need to bring about a regional planning perspective in the overall planning processes.

Evidently, the domestic and industrial water needs are a priority as enunciated in the Water Policy, 2002 (Ministry of Water Resources 2002). In the course of making provisions for these sectors, there seems to be a lack of provisioning for the handling of waste water. How do we dispose off the waste waters to the lower reaches, especially if the cities are also industrial centers? The quantity as well as quality of water supplied to these areas is of great concern and, therefore, it is important to discover whether any improvements could be made in the existing arrangements that determine the use of resources as well as consider a holistic plan that addresses planning components on a regional scale.

Acknowledgements

We wish to thank the First Year students of the 2009–11 batch of Industrial Area Planning and Management and of the Infrastructure Planning of Faculty of Planning and Public Policy, CEPT University, Ahmedabad, on whose interactive workshop (2009–10) project some of the analyses of this chapter has been based. We also wish to extend our sincere thanks to Ms Rutuja and Mr Ankit, students of the Faculty of Planning and Public Policy, for their help in accessing some of the data and analysis; and to the anonymous referee and SaciWATERS, Hyderabad, India, for their helpful comments and suggestions.

Notes

1. In the Census of India, 2001 (Government of India 2001), the definition of urban area adopted is as follows: (*a*) All statutory places with a municipality, corporation, cantonment board, or notified town area committee, etc. and (*b*) A place satisfying the following three criteria simultaneously:
   (*i*) a minimum population of 5,000;
   (*ii*) at least 75 percent of the male working population engaged in non-agricultural pursuits; and
   (*iii*) a density of population of at least 400 per sq. km. (1,000 per sq. mile).
   An Urban Agglomeration is a continuous urban spread constituting a town and its adjoining urban outgrowths (OGs), or two or more

physically contiguous towns together and any adjoining urban outgrowths of such towns. An Urban Agglomeration would be constituted as: (1.) a city or town with one or more contiguous outgrowths; (2.) two or more adjoining towns with their outgrowths; and (3.) a city and one or more adjoining towns with their outgrowths all of which form a continuous spread (Government of India 2001). Towns with population of 1,00,000 and above are called cities (ibid.).

2. It is the major industrial corridor of the country that comprises Mumbai, Nashik, Surat, Vadodara, and Ahmedabad.

3. Designated best use of water: "A"–Drinking water source without treatment but after disinfection; "B"–Outdoor bathing; "C"–Drinking water source after conventional treatment and disinfection; "D"–Propagation of wildlife and fisheries; and "E"–Irrigation, industrial cooling, controlled waste disposal (see Central Pollution Control Board and Winrock International India 2006).

References

Amarasinghe, U. A., Bharat R. Sharma, Noel Aloysius, Christopher Scott, Vladimir Smakhtin, and Charlotte de Fraiture. 2004. *Spatial Variation in Water Supply and Demand across River Basins of India*. Research Report 83. Colombo, Sri Lanka: International Water Management Institute (IWMI).

Centre for Environment Planning and Technology (CEPT). 2006a. *Ahmedabad City Development Plan 2006–2012*. Ahmedabad: CEPT University.

———. 2006b. *Surat City Development Plan 2006–2012*. Ahmedabad: CEPT University.

———. 2009. *CEPT Interactive Workshop Report 2009–10: Phase 1.* Unpublished. Ahmedabad: CEPT University.

———. 2009. *Study of Rural Livelihood along the Bank of Sabarmati River*. Rural Component Lab Report 2009. Submitted by students of Infrastructure Planning and Industrial Area Planning and Management, Faculty of Planning and Public Policy. Ahmedabad: CEPT University.

Drechsel, Pay, Christopher Scott, Liqa A. Raschid-Sally, Mark Redwood, and Akiça Bahri. 2009. *Waste Water Irrigation and Health, Assessing and Mitigating Risk in Low Income Countries*, London: IWMI and Earthscan Publication.

Gopalakrishnan, M. n.d. "An Integrated Water Assessment Model for Future Scenario Studies of Sabarmati River Basin in India." Available at http://www.iwmi.cgiar.org/Research_Impacts/Research_Themes/BasinWaterManagement/RIPARWIN/PDFs/32_Gopalakrishnan_SS_FINAL_EDIT[1].pdf (accessed August 8, 2010).

Government of Gujarat. 2006. *Census of India 2001.* Series 25. Gujarat District Census Handbook, Village and Town Directory, Surat district. Gandhinagar: Gujarat Census Operaitons.

Government of India. 2001. *Census.* Ministry of Home Affairs. Available at http://www.pon.nic.in/open/depts/ecostat/census/homepage/htm (accessed May 14, 2003).

———. 2002. *National Water Policy.* Ministry of Water Resources. Available at http://mowr.gov.in/writereaddata/linkimages/nwp20025617515534. pdf (accessed August 8, 2010).

Parthasarathy, R. (with Jharna Pathak). 2009. *State Water Sector Interventions in Gujarat: Current Status, Emerging Issues and Needed Strategies.* A report submitted to IWMI. Colombo: IWMI.

Parthasarathy, R. and Soumini Raja. 2009. "Coastal Livelihood Vulnerability to Changing Climate: A Governance Perspective." Paper presented at the 10th International Congress of Asian Planning Schools Association– "Future of Asian Cities", November 24–26, 2009, CEPT University, Ahmedabad.

Winrock International India. 2006. *Urban Wastewater: Livelihoods, Health and Environmental impacts in India.* Research report submitted to Comprehensive Assessment of Water Management in Agriculture, Colombo. Available at http://www.iwmi.cgiar.org/Assessment/ files_new/research_projects/Urban%20Wastewater-Full_Report.pdf. (accessed August 8, 2010).

14

Water Management in Rapidly Urbanizing Kathmandu Valley

Balancing Structural Linkages among Water, Society and Settlement

Bijaya K. Shrestha and *Sushmita Shrestha*

Throughout the transformation of cities of the Kathmandu Valley–from the prehistoric settlement called *grama* (of the "Kirata" period) to the commercial center *dranga*, during the Lichchhavi era (1st–9th centuries), to medieval towns at the time of the Malla dynasty (13th–18th centuries) (Regmi 1965; Oldfield 1974; Malla 1978; Slusser 1982)–water has shaped settlement patterns and influenced the daily lives of the inhabitants. Conservation of various water resources and fulfilment of water, needed for multiple activities, was successfully managed in those periods. However, rapid urbanization of the Valley together with a gradual shift in the economic base, from agriculture to services and commerce including changing lifestyles of inhabitants, has not only affected the earlier water infrastructure network but also amplified the problems of high water demand and waste water treatment. It has also transcended the management capacity of the government and has become a big challenge to those working in water management. Against this background, this chapter aims to explore the water management system in the rapidly urbanizing Kathmandu Valley, focussing on structural linkages among water, society and settlement.

Water Infrastructure and its Management in the Lichchhavi and Malla Periods

Water has been associated with settlement and society in the Valley throughout history. The skill and knowledge used by the Kiratas, (prehistoric rulers) for developing human settlements in the *tar* (elevated land), leaving fertile river flood plains for agriculture, was enhanced during the Lichchhavi period with the development of extensive water-related infrastructure, such as canals, ponds, wells, and their distribution and drainage system, which were further extended and maintained by the Malla kings (Arnstroem 1994; Becker-Ritterspach 1994; Joshi 1993; Theophile and Joshi 1992; Tiwari 2001, 2002). Numerous water infrastructure (along with their relationship with settlement and society), which developed over different cultural periods, can be categorized into three hierarchies at different scales (see Table 14.1).

At the neighborhood scale, watering places in the form of sunken stone spouts (*dhunge-dhara* or *hiti* in the Newari language), wells and drinking fountains had been scattered within the tight urban fabric of the Valley. They were not only a venue for socio-cultural activities of the "Newar" community (Pradhan 1990) but also the architectural, artistic, social, and engineering heritage of

Table 14.1: Structural linkage between water, society and settlement in the Kathmandu Valley

	Linkage	Neighborhood level	Local level	Town level
Water	Society	Socio–religious activities in daily life	Production and distribution	Activity associated with birth and death
	Settlement	Water conduits, wells, ponds, etc.	Canals, natural ponds, reservoirs, etc.	River & ponds as town boundary

Source: Prepared by the authors.

these ancient people (Spodek 2002). Water conduits along with *paati* (rest place) and *sattal* (rest shelter) were provided at three different locations to serve a wide range of beneficiaries: (*a*) at the neighborhood areas for local inhabitants, (*b*) outside the settlement for travellers, and (*c*) at the foothills of the mountain for pilgrims (Becker-Ritterspach 1995). Among the numerous infrastructural extensions during the Malla period, the long distance water canals, known as *rajkulos* (starting from the mountain foothills) and their distribution systems were significant. They were used for religious and irrigation purposes as well as for feeding historical ponds, whose function was to recharge the shallow aquifer (see Table 14.2). A buffer zone of agricultural land existed between the settlement and the rivers. The waste water of the town was used for agricultural purposes before reaching the rivers. Irrigation also helped maintain groundwater levels.

As water is essential for different rituals (from birth until death), many artificial ponds, such as *kamal pokhari* and *siddha pokhari*, during the amalgamation of Lichchhavi settlements in Bhaktapur, and *dev, kha* and *po* ponds, at the time of forming new settlements in Bungamati, were constructed for ritual bathing. Such artificial ponds had also a function of rainwater harvesting and storing of rainwater, which was used to recharge groundwater and ultimately fed into stone spouts and wells. Major rivers at the town periphery also acted as a transitional space between the two worlds–human habitats, inside the town, and the domain of the death, outside the river. They were not only sites for major temple complexes (for example: Pashupatinath and Shova Bhagawati in Kathmandu) but also the locus for dying, cremation and purification.

Table 14.2: Development of water system during the Lichchhavi and Malla periods

Towns	Irrigation canal	No. of historical ponds	No. of ponds served by *rajkulo*	No. of shallow aquifers served by *rajkulo*
Kathmandu	*Rajkulo*	21	4	7
Patan	*Rajkulo*	39	18	11
Bhaktapur	*Rajkulo*	30	9	4

Source: Shrestha and Shrestha 2008.

Cultural tradition, tangible and intangible–in the form of celebration of festivals and rituals with wide community participation–as well as the inhabitants' religious beliefs, social norms and spiritual values, causing attitudinal and behavioural changes, had helped to manage water infrastructure at different scales during the Lichchhavi and Malla periods. First, polluting water sources and damaging water infrastructure never even crossed people's minds. Individual people considered the *naaga* (serpent) as a source of water and believed that anyone who agitated the *naaga*, by polluting the water sources, would suffer from skin diseases and infections. Offerings were made at the gutter hole (the point where the gutter crosses the boundary of the plot) of the *saagah* (dumping site), considered to be the dwelling of the *naaga*, when someone became ill. Second, *dhunge-dharas* and wells were the lifelines of the communities and part of daily rituals. Sharing the same facilities had strengthened community networks and social support. As water conduits comprised both Hindu and Buddhist pantheons (symbolizing their holiness), people believed that by taking a bath in *dhunge-dhara*, one would get religious merit equal to visiting all the important *tirthas* (holy places) of both religions. Third, establishing water spout and other community facilities was considered a pious act in the "Newar" culture. It was sporadic, individually initiated and motivated by benevolent intentions. Most of these were associated with the community, unlike the temples and shrines established by royal families.

The two institutions for operation, maintenance and management of canals and other water infrastructure, at the local level, were the *guthi* system and celebration of annual festivals. *Guthi*, a corporate body financed to perpetuity through land grants or other fixed deposits, was created either by an individual or by the government for continuation of community services and religious activity over many generations. Different occasions were used for the *puja* (worship) and maintenance of *dhunge-dhara* through mobilization of whole neighborhood under the *guthi*'s leadership. Annual maintenance of these public utilities, to ensure the continuous flow of water even in dry season, has been achieved through celebration of *sithinakha* festival, dedicated to the ancestors (*digu* puja or Dewali) in the month of May, by repairing public buildings and urban services such as wells, water holes, ponds, and drainage ditches through wide community participation.

Large water system at the town level was, however, managed through celebration of annual festivals with the involvement of state and community. First, the "Rato Matsyendranath Jatra" (festival related to water, dedicated to the rain goddess) was initiated to ensure the functioning of the Tikabhairav *rajkulo* and maintaining numerous ponds and stone spouts within Patan. Before constructing the *ratha* (chariot) for performing the *jatra* (journey), there had to be water in all the major ponds charged by the *rajkulo*. Similarly, the cane used for making the *ratha* had to be wetted at the *la pukhu* (water pond) of Pulchowk, whereas materials for preparation of *bau pichaa* (sacred vessel made of bamboo, which is used for offering food items, such as rice, grains and vegetables, to the gods and goddesses) had to be sunk by twelve *nayos* (leaders) of "Karunamaya" at *thapa hiti* (sunken water spout). The major maintenance used to be carried out every 12 years as a great event with the whole community visiting Bungamati.

Second, people from various *toles* (neighborhoods) and different castes were given specific responsibilities in the *jatra*, which not only helped to unite them but also developed a sense of belonging and ownership. Wide community participation was achieved by inviting people from Bhaktapur, Thimi, Bode, Nagadesh, Kirtipur, and Panga (Locke 1980) in the construction and maintenance of this great water work.

Third, larger water bodies, such as Bagmati, Vishnumati and Sali rivers, along with major temple complexes were considered as holy places and taking a bath in such places at different occasions was considered to have many religious benefits. It was believed that the water had to be kept clean and the amenities for the pilgrims intact for celebration of annual festivals.

Consequences of Importing Neoliberal Models for Regulating Urban Transformation

These practices sustained the water management system for many centuries during the Lichchhavi and Malla periods. However, they began to fade by the end of the 18th century during the "Rana" regime due to British–European influences in town planning and architecture. The neglect continued when the planned development was initiated in the Valley during the 1950s, after the end of "Rana" autocracy. The government adopted neoliberal policies to fulfil the water needs of increasing population and to regulate the urban

244 *Bijaya K. Shrestha* and *Sushmita Shrestha*

growth. Formulation of a series of acts and legislation, together with the new institutional set-up, to implement the various development plans on linear basis was found to be ineffective to address the complex nature of numerous urban problems.

Instead of regulating urbanization, successive governments have adopted centralized policies and formulated liberal economic policies after 1991. The enactment of the Industrial Enterprise Act, 1992, allowing the private sector to establish, expand and upgrade facilities (except those related to defence, public health and environment) without a license, together with tax exemption and rebates has accelerated the growth of many industries at Balaju, Patan and Bhaktapur as well as along the Koteshwar–Bhaktapur and Kalanki–Thankot highway corridors. Similarly, the government's shift in policy from "housing provider" (in the 1970s) to "housing facilitator," encouraging private developers (in the 1990s) has further accelerated the development of high-rise apartments, particularly in the northern part of the Valley. Rapid construction of industrial and service facilities and new high-rise apartments, including tourist-related infrastructure, have dramatically changed the landscape of Kathmandu (Kobayashi 2006). The overall consequences are twofold: (*a*) the destruction of traditional water infrastructure and their interdependence, and (*b*) inadequate and ineffective responses to water demand in the processes of socio-economic modernization of the Valley.

Destruction of Traditional Water Infrastructure and Its Management System

The destruction of traditional water infrastructure has taken place in three different phases. Unlike the Malla kings, the *khas* rulers (that included the Shah and Ranas) from the hills had poor knowledge in culture, science and technology and were also remote from the common people (UN Habitat 2008). Hence, instead of continuing the indigenous water resource management, which had been sustained for many centuries, they adopted the municipal water supply system by constructing pipelines and public stand posts in several places from the 1960s. Water from ponds and aquifers was also diverted into the fountains and gardens of the "Rana" palaces, leaving the traditional stone spouts and wells unattended and unmanaged, marking the first stage of destruction. Another major setback (in the traditional practice) occurred when the government decided to manage the town's religious faiths and community services through a semi-government *guthi* corporation. All the properties from private *guthis* were taken over by the corporation and a new policy of replacing that land with cash was formulated. From the mid-1970s, public agencies along with community institutions, clubs and neighborhood organizations started destroying, encroaching upon and damaging numerous ecological sites, such as natural aquifers, ponds and their distribution systems, by constructing buildings over them. Haphazard laying down of underground sewer lines (after 1978) further damaged the flow path of aquifers, obstructed the natural flow of water and drainage, and mixed the water with sewage. Construction of basements in the new buildings, in general, and digging up of shallow wells by individuals, in particular, over the period of 1970–90, has also significantly decreased the discharge of spouts. The final stage of destruction accelerated after 1991 due to the indiscriminate extension of settlements towards the riversides, encroachment of riverbanks by slums and squatter settlements, dumping of garbage, and direct discharge of sewer lines and industrial waste in to the river without treatment, thereby converting the holy rivers and their tributaries into open sewer lines.

Impacts on Water Resources in the Modern Period

The impacts of increase in population density and their socio-economic activities, including adaptation of modern lifestyle, on water resource management at present are numerous. First, there is a huge gap between the high demand for and low supply of water in the Valley, where the supply is just half of the total demand. While the daily water demand is 294 million liters per day (mld), the average daily production is just 145 mld (NWSC 2004); and leakage accounts for about 50 percent of the total distribution due to the old pipe network (Annapurna Post 2009). The existing capacity of the Nepal Water Supply Corporation (NWSC)–comprising nine major supply systems, 15 water treatment plants and 132,803 legal connections, including 809 community taps (NGOFUWS 2005)–is able to cover 70 percent of the total urban areas of the Valley with only 47.4 percent of the households having a connection (NWSC 2001). Out of 14 mineral water industries, nine are located in the Valley, which supply about 0.9 mld per day (Dol 2006). Low income groups (LIGs) are able to consume only 10 liters of water per day, which is insufficient to maintain basic hygiene and sanitation standards. Many people still depend on stone spouts, especially in Patan and Bhaktapur. Piped water supply from NWSC is unreliable, irregular and of low discharge, and polluted in most cases. Water is supplied for only one–two hours once a week in the dry season. Households need to get up in the early hours of the morning to extract water from the main line using electric pumps. Others have to purchase it from private vendors at high price. They engage in five main types of coping behaviours–collecting, pumping, treating, storing, and purchasing (Pattanayak et. al 2005). Boiling and filtering, including use of "Euro Guard"[1] and Solar Disinfection System (SODIS),[2] are the different methods used for purifying water. Even then, many do not treat water.

Second, numerous activities such as dumping debris and agricultural residue, cremation of dead bodies, urban solid waste disposal, household effluents and open defecation, slaughtering, squatter settlements, animal sheds, washing clothes, and quarrying of pebbles and sand are also polluting the rivers, ponds and other sources of water. The Valley alone hosts more than 72 percent of the country's water polluting industries. Although several wastewater treatment plants have been constructed over the years in the Kathmandu Valley, none, except for one at Gaurighat and another at Thimi, are functional. In a study, 82.6 percent and 92.4 percent of drinking water samples were found to cross the World Health Organization (WHO) guideline values for total plate and coliform count for drinking water (Prasai et al. 2007). Deep tube well water is contaminated by arsenic (JICA-ENPHO 2005). The geochemical study of fluvio-lacustrine aquifers of the Kathmandu Valley found high levels of arsenic (Gurung et al. 2007) that can cause cancer.

Third, use of polluted water has negative consequences for public health and results in high social cost. Admission of 1,360 diarrheal patients to the "Sukraraj Tropical Infectious Disease Hospital" between May 2 and 21, 2004 (NGOFUWS 2005) illustrates the seriousness of the incidence of water-borne diseases in the Valley. A report from "Teku Hospital" in Kathmandu reveals that 16.5 percent of all deaths had been due to water-borne diseases (Metcalf & Eddy 2000).Water-related diseases, such as diarrhoea, stomach ailments, and dysentery, have also impacted significantly on productivity and income because of the loss in working days and treatment expenses.

Fourth, as municipal supply of water is poorly managed and service provided is inadequate, large quantities of groundwater are pumped by both municipal authorities and private individuals, to meet domestic and other water needs. In addition, most industries, hotels and corporate houses pump water from deeper aquifers. The total sustainable withdrawal of groundwater from the Valley's aquifers is 26.3 mld (Stanley et al. 1994). However, as high as 58.6 mld of groundwater has already been extracted (Metcalf & Eddy, 2000), 8.4 percent from shallow wells and 91.6 percent from deep wells (KMC & World Bank 2001)–thereby causing "groundwater mining."[3] Groundwater levels in these aquifers have dropped from 9 m to as low as 68 m over a few years. Rapid expansion of human settlements and conversion of forest lands into agriculture plots have also constrained the groundwater recharge from the surrounding watershed areas. Extensive sand mining from the riverbeds is contributing to lowering the water table of adjoining fields, thereby adversely affecting the aquaculture as well as increasing the cost of pumping deeper boreholes.

Fifth, there is a visible conflict at different levels and sectors both in urban and rural areas of the Valley. In urban areas, it is not only households that are unable to rent their rooms for many months but frequent quarrels among the family members are common due to lack of water. At the neighborhood level, there is often dispute regarding who should get to fill their water container first, while collecting water either from the government tanker or from the nearby stone spouts. In some cases, the local community protests at the NWSC either for dry taps or against the big hotels and industries (located in their neighborhoods) for pumping all water from the municipal lines using powerful pumps.

Though such conflicts are in early stages, the incidence of a case of a 13-year-old girl, Ramila Biswakarma, who was beaten to death over a minor dispute with a 45-year-old local resident in Letang, Morang district, on the issue of who should get to fill the water container first (Upadhay 2010) has clearly demonstrated how the issue can bring social unrest. The current trend of supplying water by the NWSC tankers to the elite and to those who have political access, thereby forcing the general public to buy expensive water from the private sector, will definitely increase such conflicts in future.

Sixth, water in the form of rivers, ponds and springs is a special type of land use and spaces around these water sources have high real estate value due to the ephemeral quality of water, which includes buoyancy, waves, currents, tides and calm, and a natural and pollution-free environment. They can also be used against fire hazard. However, such multiple potential values of water have been greatly ignored in the urban development programs and policies.

Last, but not the least, though the ambitious Melamchi project is aiming to bring about 510 mld water per day in the Valley through (*a*) the Melamchi River (first phase: 170 mld), (*b*) Yangri river (second phase: 170 mld), and (*c*) the on-going Larke river project (third phase: 170 mld) (ICIMOD et al. 2007)–it will take many years for the project to deliver water. Since the cost of the project will be recovered from the consumers as per the "*full cost recovery*" principle, as prescribed by the Asian Development Bank (ADB) in its water policy, the actual cost to be borne by the community will be much higher than the present water tariff due to the long delay in construction. Such mega projects alone will not improve the existing situation unless the distribution of pipe lines is improved, urban growth is regulated and sufficient waste treatment plants are constructed.

Legal and Institutional Framework

The neoliberal model adopted for water management in Nepal after the end of the "Rana" period was a new concept for the government authorities as well as the people. It relies on establishing new institutions, policies and acts. Unlike the traditional water management system, it is a "top-down" approach requiring co-ordination among different agencies at the central and sectoral levels and clear-cut policies and acts. Moreover, such a model has given little attention to giving a sense of ownership to the community. Besides, it depends more on scientific data rather than the users' attitude and behaviour. As a result, despite the introduction of several institutions (at the central level supported by various acts), scarcity and conflict over water still prevails in the Valley.

The reasons are numerous. First, the establishment of new institutions and enactment of various acts is quite influenced by international practices, rather than being the product of an actual need analysis at the local level. For instance, many local communities are not familiar with Water Users Association (WUA) and Water Users Groups (WUGs), which are administered either by the District Development Committee (DDC) or Village Development

Committee (VDC) while some do not feel comfortable in these political institutions. In other cases, international declarations concerning water, such as the International Conference on Water and the Environment in Dublin in 1992 (Hydrology and Water Resources Department, World Meteorological Organization 1992), the Rio Summit in 1992 (UN Conference on Environment and Development 1992) and the Second World Water Forum in 2000, including the Ramsar Convention on Wetlands, 1971, have influenced the formulation of environmental protection laws and water resource strategies as well as shaped donor-funded water projects. At present, the Ministry of Water Resources (responsible for hydropower, irrigation and disaster prevention) and the Ministry of Physical Planning and Works (responsible for drinking water supply) are two major central-level institutions. Various departments under these two ministries, together with other departments belonging to different ministries, are acting as sub-sectoral implementing agencies. However, none of these agencies and contemporary acts deal with the multiple functions of water and its linkages with society and settlements at the different levels. In fact, each sub-sector is looking for single use of water with little co-ordination and co-operation with other sub-sectors working on water management. Fragmentation of management practices, among many government agencies and semi-government organizations, has resulted in a failure to consider the cross-sectoral effects of water activities leading to waste and sub-optimal allocation. As a result, these acts and strategies have become more like a list of recommendations with good aims and objectives but with little scope of implementation.

Second, contemporary acts have failed to change the community's attitude and behaviour towards consumption patterns and create a sense of belonging. None of them acknowledge the religious value of water—one of the most important aspects of traditional water management. The present approach of using the community for labour and material without involving them in the decision-making process, particularly in small-scale drinking water projects, has created the habit of the community depending on donors for construction of water taps with little responsibility towards operation and maintenance. As the local governments have been functioning without elected mayors for the last few years, due to political instability, most of the WUGs are also ineffective and defunct while the newer programs hardly include the users. Consequently, the opportunity of reducing the environmental and social costs using local and indigenous knowledge and consultation has been lost.

Third, the managerial, technical and financial capabilities of public agencies working on urban development and in the water sector are poor; these agencies have overlapping and conflicting duties and responsibilities. Such confusion is also apparent in the existing (different) acts. Despite the presence of five layers of public structures for the planned development of the Valley, authorities are yet to prepare a comprehensive master plan with the formulation of land use zoning, planning standard and urban design guidelines. The Nepal Water Supply Corporation Act, 1989, empowers the corporation to set and enforce water fees, prosecute illegal connections and prevent the misuse of drinking water, including the responsibility for providing urban water and sanitation, controlling water pollution, and the maintenance of urban sewerage systems and treatment plants. Similar responsibility on water and sanitation (including the construction of treatment plants) has also been given to local municipalities by the Local Self Governance Act 1999. The former is under the jurisdiction of the Ministry of Physical Planning and Works whereas the latter works under the Ministry of Local Development. Though households of the Valley are willing to pay for safe and adequate water (Whittington et al. 2002; Tiwari 2000), the NWSC, supported by a number of development partners for more than 30 years, has neither been able to deliver efficient and affordable services on a sustainable basis, nor significantly expand its service coverage, due to excessive political intervention, inadequate and inefficient operation and maintenance, and weak revenue collection. The public holds the NWSC in low esteem and does not believe that it is working in their best interest. The politicization of water has resulted in the practice of heavy dependence on government agencies for developing, operating and maintaining water systems, with noticeable absence of incentives for profitability and efficiency that typically motivate market participants.

Fourth, water infrastructure developed in the Lichchhavi and Malla periods, together with the different festivals and rituals, is a national heritage to be conserved through policy and legislation and by marking them as special land use in the city plans. Though the Environmental Protection Act, 1997, provides legal basis for declaring the traditional water system as heritage, it has not yet been realized in actuality.

Fifth, both public and non-government organizations are yet to show any signs of seriousness in dealing with the emerging issue of water conflict between different sectors and communities.

Conclusion and Recommendations

The present sorry state of water infrastructure of the Kathmandu Valley is not only due to rapid urbanization, uncontrolled population growth and chaotic physical infrastructure development, but also because of myopic planning, poor policy environment and mismanagement of the existing water resources. The success of the traditional water resource management rested on three interrelated aspects, namely, (*a*) the innovative and interlinked water infrastructure network at three different levels, (*b*) the community's religious beliefs, social norms and spiritual values, and (*c*) the social institution of the *guthi* system and celebration of various festivals. Thus, a mechanism linking "water, society and settlement" was able to conserve water resources and fulfil the community's multiple needs for many centuries during the Lichchhavi and Malla periods. Instead of continuing this unique community-based management system, the Shahand Rana rulers introduced a new system of municipal pipe lines. The demise of the *guthi* system has disintegrated community network, social support and a sense of ownership over water infrastructure, while random installation of public utilities and construction of non-engineering buildings has destroyed the *rajkulos* and its distribution system. Failure to regulate rapid urbanization, socio-economic modernization and changes in people's lifestyles (since the last two decades) has not only further ruined the traditional water infrastructure but also created water stress. The imported neoliberal model, backed by the new institutional set-up and series of water and environmental acts, is simply ineffective to address these numerous emerging problems. In such a situation, a combination of both "top-down" approach–by integrating water with rapid urban growth and society in the legal and institutional framework, at the strategic level–and "bottom-up" technique–by revitalizing the traditional water infrastructure as well as educating households for water conservation is essential. To achieve these goals, the following recommendations will be necessary:

(a) Regulate the present trend of haphazard urban growth by decentralizing business activities in the peripheral areas as well as outside the Valley, and use urban design guidelines and economic incentives for desirable planning and design of settlement and building construction, so that water can be equally distributed and, at the same time, minimize the pollution of rivers and water springs.

(b) Divide the Valley into different zones, based on the existing water infrastructure situation, population density and business activities, and accordingly develop the plans and programs for water supply and demand, reducing leakage from the pipelines and improving waste water management.

(c) Develop water harvesting system at city and household levels to take the advantage of 1,600 mm average annual rainfall. Even storage of 80 percent rainwater in an area of 100 sq. m. will give 128,000 liters of water annually, which is enough for conducting water-related tasks at the household level. It will also encourage households to use toilet and bathroom fixtures that need little water to clean.

(d) Finally, enhance the traditional practices, local festivals and rituals as well as revitalize the traditional water infrastructure.

Notes

1. It is a portable, household vessel, used for purifying drinking water and needs only 25 watts of electricity. This vessel filters the water in three different stages–pre-filter (to remove impurities such as bacteria, dust, dirt, and mud); activated carbon block (to reduce organic impurities such as bad color, bad smell, bad taste, and free chemicals such as chlorine, ammonia and lead); and ultra violet chamber (to kill harmful microbes).

2. It consists of filling plastic bottles with water and exposing them to sunlight to disinfect bacteria and viruses–which requires at least 500 W/m of radiation for five hours.

3. The process of extracting groundwater from a source at a rate that exceeds the replenishment rate such that the groundwater declines persistently; thereby, not only reducing water levels in the wells, ponds and aquifers but also increasing the threats of landslides and the vulnerability of the settlements (buildings and other infrastructure).

References

Annapurna Post. 2009. "Fifty Percent Leakage in Valley's Drinking Water System," January 19. Available at http://www.ngoforum.net/index. php?option=com_content&task=view&id=4854 (accessed April 1, 2013).

Arnstroem, E. K. V. 1994. "Water Supply in Lalitpur, Nepal–Case Study: Storage and Redistribution of Water from Hities (Stone Spouts) in a Small, Urban Area." A Thesis, Lulea University of Technology, Sweden.

Becker-Ritterspach, R.O. A. 1995. *Water Conduits in the Kathmandu Valley.* New Delhi: Munshilal Manoharlal.

Department of Industry (DoI). 2006. *Production and Capacity of Mineral Water Industries Registered* (computer records). Unpublished paper. Kathmandu: Department of Industry.

Government of Nepal. 1999. *Local Self-Governance Act.* Available at http://www.lawcommission.gov.np/en/documents/prevailing-laws/prevailing-acts/Prevailing-Laws/Statutes---Acts/English/Local-Self-governance-Act-2055-%281999%29/ (accessed March 1, 2010).

Gurung, J. K., I. Hiroaki, M. H. Khadka, and N. R. Shrestha. 2007. "The Geochemical Study of Fluvio-Lacustrine Aquifers in the Kathmandu Basin (Nepal) and the Implications for the Modernization of Arsenic", *Environmental Geology*, 52(3): 503–17.

Hydrology and Water Resources Department. 1992. The Dublin Statement and Report of the Conference," in Proceedings of *International Conference on Water and the Environment: Development Issues for the 21st Century*, January 26–31, World Meteorological Organization, Dublin, Ireland.

ICIMOD (International Centre for Integrated Mountain Development), MOEST (Ministry of Environment, Science and Technology) and UNEP (United Nations Environment Program). 2007. *Kathmandu Valley Environment Outlook.* Kathmandu: ICIMOD.

JICA (Japan International Cooperation Agency)–ENPHO (Environment and Public Health Organisation). 2005. *Groundwater Quality Surveillance in Kathmandu and Lalitpur Municipality Areas.* Kathmandu: JICA–ENPHO.

Joshi, P. R. 1993. *Feasibility Study of Rajkulo: Rehabilitation of Patan's Traditional Water Supply Network.* Final unpublished report. Patan Conservation and Development Programme, GTZ–Urban Development through Local Efforts Programme (UDLE).

KMC (Kathmandu Metropolitan City) and World Bank. 2001. *City Diagnostic Report for City Development Strategy.* Kathmandu.

Kobayashi, M. 2006. "Social Change in Kathmandu Related with Globalisation and Liberalisation: Potential of New Life Style and Domestic Market," *Toyo University Journal*, 44(1): 27–38.

Locke, J. K. 1980. *Karunamaya: The Cult of Avalokitesvara-Matsyendranath in the Valley of Nepal.* The Research Centre for Nepal and Asian Studies, Tribhuvan University, Kathmandu: Sahayogi Prakashan.

Malla, U.M. 1978. "Settlement Geography of Kathmandu Valley," *Geographical Journal of Nepal*, 1(1): 28–36.

Metcalf & Eddy Inc. 2000. "Groundwater and Wastewater." Paper presented at a seminar on Ground Water and Waste Water, February 14, organized by the Melamchi Water Supply Development Board, Kathmandu.

NGO Forum for Urban Water and Sanitation (NGOFUWS). 2005. *Delivering Water to the Poor: A Case Study of The Kathmandu Valley Urban Water Supply Reforms with a Special Focus on the Melamchi Project.* Kathmandu: NGOFUWS. Available at http://www.ngoforum.net/index.php?option=com_docman&task=doc_view&gid=14 (accessed April 1, 2013).

Nepal Gazette. 1997. "Environment Protection Act," June 24. Available at http://www.lawcommission.gov.np/en/documents/prevailing-laws/prevailing-acts/func-startdown/9/ (accessed March 1, 2010).

Nepal Water Supply Corporation. 1989. *Nepal Water Supply Corporation Act.* Available at http://www.lawcommission.gov.np/en/documents/prevailing-laws/prevailing-acts/Prevailing-Laws/Statutes---Acts/English/Nepal-Water-Supply-Corporation-Act-2045-%281989%29/ (accessed March 1, 2013).

———. 2001. *Annual Report 2001.* Kathmandu: NWSC.

———. 2004. *Annual Report 2004.* NWSC: Kathmandu.

Oldfield, H. A. 1974. *Sketches from Nepal.* Delhi: Cosmo Publication.

Pattanayak, S. K., J. C. Yang, D. Whittington, and K. C. Bal Kumar. 2005. "Coping with Unreliable Public Water Supplies: Averting Expenditures by Households in Kathmandu, Nepal", *Water Resources Research*, 41(2), (W02012), doi:10.1029/2003WR002443.

Prasai, T., B. Lekhak, D. R. Joshi, and M. P. Baral. 2007. "Microbiological Analysis of Drinking Water of Kathmandu Valley," *Scientific World*, 5(5): 112–14.

Pradhan, R. 1990. "Dhunge–Dhara: A Case Study of the Three Cities of Kathmandu Valley," *Ancient Nepal*, 116–118: 10–14.

Ramsar Convention on Wetlands. 1971. "The Convention on Welands Text, as Amended in 1982 and 1987–Convention on Wetlands of International Importance especially as Waterfowl Habitat." Available at http://www.ramsar.org/cda/en/ramsar-documents-texts-convention-on/main/ramsar/1-31-38%5E20671_4000_0__ (accessed March 3, 2010).

Regmi, D. R. 1965. *Medieval Nepal, Part 1(Early Medieval Period, 750–1350 A.D.).* Calcutta: Firma K. L. Mukhopadhyay.

Second World Water Forum. 2000. "Ministerial Declaration of The Hague on Water Security in the 21st Century." Available at http://www.waternunc.com/gb/secwwf12.htm (accessed March 3, 2010).

Shrestha, S. and B. K. Shrestha. 2008. "The Influence of Water in Shaping Culture and Modernization of the Kathmandu Valley," in Jan Feyen, Kelly Shanon and Matthew Nevile (eds), *Water and Urban Development Paradigms–Towards an Integration of Engineering, Design and Management Approaches*, pp. 183–88. London: CRC Press.

Slusser, M. S. 1982. *Nepal Mandala, Volume I.* Princeton, NJ (UC): Princeton University Press.

Spodek, J. C. 2002. "Ancient Newari Water-Supply Systems in Nepal's Kathmandu Valley," *Association for Preservation Technology [APT] International Bulletin*, 33(2/3): 65–69.

Stanley International Ltd, Mott MacDonald Ltd and East Consult (P) Ltd. 1994. *Bagmati Basin Water Management Strategy and Investment Program.* Final Report. His Majesty's Government of Nepal, Kathmandu: Ministry of Housing and Physical Planning, Japan International Cooperation Agency and World Bank.

Theophile, E. and P. R. Joshi. 1992. *Historical Hiti and Pokhari: Traditional Solutions to Water Scarcity in Patan.* Unpublished report. Patan Conservation and Development Programme, UDLE/GTZ.

Tiwari, D. 2000. *Users' Willingness to Pay for Averting Environmental Health Risks and Implications for Alternative Policy Measures for Improving Water Supply Facilities in Kathmandu Valley.* Final report. Kathmandu: Department of Water Supply and Sewerage.

Tiwari, S. R. 2002. *Transforming Patan's Cultural Heritage into Sustainable Future: Case Studies of the Past and the Present.* Unpublished paper.

UN Conference on Environment and Development. 1992. "Rio Earth Summit." Available at http://www.un.org/geninfo/bp/enviro.html (accessed March 3, 2010).

UN Habitat. 2008. *Water Movement in Patan with Reference to Traditional Stone Spouts.* Kathmandu: UN-HABITAT Water for Asian Cities Programme Nepal.

Upadhay, B. 2010. "Running on Empty," *The Kathmandu Post*, March 25.

Whittington, D., S. K. Pattanayak, J. C. Yang, and K. C. Bal Kumar. 2002. "Household Demand for Improved Water Services: Evidence from Kathmandu Nepal," *Water Policy*, 4(6): 531–56.

15

Private Water Tanker Operators in Kathmandu

Analysis of Water Services and Regulatory Provisions

Dibesh Shrestha and
Ashutosh Shukla

—

Kathmandu is growing and so is its water requirement.
Historically, Kathmandu has a rich tradition of community-based
water supply schemes that have conventionally supported the
drinking water needs of the city. King Bhavari, grandson of King
Mandev I, constructed one of the first water conduits in 550 CE
in Hadigoan (Vajracharya 1973; Pradhan 2003). Many such water
conduits, small drinking water tanks, stone spouts, dug-wells and
ponds, constructed in the different historical periods, are still in
operation and provide water to the residents of the city. In the
late 1980s, to address the growing water needs and to regulate
piped water supply in Kathmandu Valley, the Nepal Water Supply
Corporation (NWSC) was established. In February 2008, the
Kathmandu Upatyaka Khanepani Limited (KUKL), took over and
became the main water service-providing agency responsible for

maintaining the piped water schemes and delivering water to the residents in different parts of the city.

The Kathmandu Valley has been undergoing rapid urbanization since 1960. There has been unprecedented growth in the population over the past four decades, especially after 1990. The population of the Valley increased from 1,105,379 (in 1991) to 1,645,091 (in 2001), with an average annual growth rate of 4.06 percent. During the same period, the urban population of Kathmandu increased from 598,528 to 995,966, with the annual growth rate of 5.22 percent; in 1952–54, only 47.4 percent of Valley's population was urban, which increased to 60.5 percent in 2001 (CBS 2003). This process of urbanization has created the demand for drinking water, increasing pressure on the surface and groundwater sources. At present, people in Kathmandu are meeting their water requirements from multiple sources that include:

(*a*) piped water supply provided by the KUKL through private water connections and public stand posts and tanker water supply (yellow tankers), in those areas where piped supply has been unreliable (Note: These tankers are operated by KUKL and they are not private in nature);

(*b*) traditional stone spouts, dug wells and ponds;

(*c*) different forms of the water market–private water tanker trucks that have been selling water (to the residents in those parts of the city where piped supply has been unreliable and/ or inadequate during periods of water shortage); packaged drinking water in the form of water jars and bottles; water vending (at kiosks) and

(*d*) groundwater extraction at domestic and industrial (hotels, manufacturing and construction industries) units to meet full or part of the water demand.

Growing Water Market in Kathmandu

Amidst the increasing water demand, the water market has created a prominent niche among the different sources of water supply in Kathmandu. Like other growing cities of South Asia, such as Chennai in India, the water market in Kathmandu relies on water transfers from rural to urban areas (Moench and Janakarajan 2006). In the face of accelerating urbanization in Kathmandu, rural–urban water transfers, in the form of water markets, are expanding. New market-based, profit-oriented and individual-centric water supply systems have emerged in numerous parts of the Valley, which vary in the scale of water extraction and magnitude of business in terms of financial transaction. The simplest form of these systems includes selling of water (obtained from dug wells) in buckets for [1]NRS 5 to 20 (US$ 0.071 to 0.286) per 20-liter-capacity bucket. Another form is extracting water, using an expensive boring system by an individual or community, followed by treating it with simple procedures and then selling it in buckets with varying charges for local communities. Such phenomena are becoming common within the communities of the Valley, mainly in Lalitpur. Yet another form is the selling of water from dug wells on an hourly basis; for instance, in the Purano-Naikap Village Development Community (VDC), individuals with *pani-ko-mul* (water source) sell water, charging NRS 800 (US$ 11.43) for supplying water for 24 hours. Water tankers along the streets of Kathmandu are another common sight. Individual water entrepreneurs install shallow or deep boring systems or use local streams for water extraction with high investments, about NRS 3,000,000 (US$ 42,857) for deep boring of 100 meters or more; they also operate the tanker business or sell water to other operators. These forms of water transactions are prevalent in Chobar, Matatirtha, Jorpati, and many other rural and urban locations of Kathmandu. The water market has different levels. For instance, in Tamsipakha, water vendors buy the water from tankers, store it in 5,000-liter tank and resell it in buckets at NRS 8 (US$ 0.114) for 20 liters capacity bucket or jar. The most complex form of water market (observed in Kathmandu) is the sale of water in sealed water bottles and jars–the bottles often come in 1 liter volume and cost NRS 15 (US$ 0.214) while a single water jar of 20 liter costs NRS 50 (US$ 0.714).

The basic reason for the expansion of such water markets in Kathmandu is the shortage of water for use because of increasing water demand due to urbanization and population growth and the limited capacity (in quality and quantity) of traditional and private sources to meet the demand. The continued failure of the water service-providing agency to meet the ever-rising demands is another factor. Moench and Janakarajan (2006) argue that this emergence of the water market is a result of the demand for a convenient water supply, created by the gap left by combined services of traditional sources and piped water supply system in Kathmandu Valley.

This chapter attempts to provide a picture of such water transfers in Kathmandu, focussing on the services provided by the private water tanker operators in terms of quantity and quality. Among the different forms of rural–urban water transfers, private water tanker operation has been the prominent mode. This market already commands a significant volume of water business, financial transactions and the engagement of different actors, from the point of extraction of water at the source to its transportation and subsequent distribution in different parts of the city. The volume of tanker-based water business has been growing with time and will grow at more accelerated rates, in the days to come, as water becomes scarcer.

However, the documentation of such water business is rare–the reason being that it is not yet been officially recognized as one of the major contributors of water supply, notwithstanding its significance (Gyawali 1988; Moench and Janakarajan 2006). This chapter is based on a study carried out by the authors during the period of August to December, 2009, to analyze the water market relating to private water tanker operation in Kathmandu. The study focussed on identifying the actors involved in this water market, the volume of business, the performance of tanker service providers, and regulatory mechanisms associated with private water tanker operation.

Methodology

The study involved semi-structured interviews with different actors involved in the water tanker operation–owner of water sources, selling water to tanker operators; tanker owners fetching water from different locations and supplying water to the residents in different parts of the city; and tanker operators (drivers and cleaners). In addition, key informant interviews were conducted with government officials involved at the policy level and those associated with the regulating agencies, officials of the KUKL and functionaries of the water tanker entrepreneurs' association.

Field visits were carried out in different locations–those used by private tankers in fetching water. The objectives of these visits were to assess source characteristics, arrangements with the source owner for water extraction, volume of water extracted on a daily basis during peak seasons (dry period from mid-January to mid-June) and off-peak season (the remaining months of the year), and methods of water treatment used. Water samples were collected from different sources to evaluate the water quality of the source; eight such samples were evaluated from five locations. It was, however, not possible to evaluate the water quality delivered to the consumers and the possible contamination in the tankers, during transportation and delivery, due to difficulty of keeping track of the tankers delivering water to different parts of the city.

Features of Tanker-Based Water Market in Kathmandu

Sources of Water and Their Locations

The tanker-based water market in Kathmandu essentially involves rural–urban water transfer including Chobar (near Taudaha), Matatirtha VDC, Jorpati, Manamaiju VDC, Gothatar, Godawari, Kakani, Baikhu, and rural areas of Bhaktapur that are close to Kathmandu. However, there are also more urbanized locations within and close to the Ring Road–which is an important [infrastructure] development in the Valley that played a crucial role in the expansion of the settlements beyond the traditional urban city core–from where the tanker entrepreneurs carry extraction, such as Balaju and the Gongabu Bus Park. The sources of water at these locations are generally spring and groundwater, though there are several other spots with spring and stream sources, occasionally tapped by the tanker entrepreneurs. The criteria used by the water entrepreneurs in selecting the sites for water extraction are ease of availability of good quality water at lower cost and distance from the source to delivery points within the city.

Tanker Truck Size

There are four sizes of water tanker trucks. For simplicity, they can be categorized into two groups: (*a*) small tanker trucks (5,000, 6,000 and 7,000 liters capacity each), and (*b*) large tanker trucks (12,000 liters capacity). The size of the tanker determines the cost of its load. The entrepreneurs operate different sizes of tanker trucks to serve the consumers' needs, according to their water needs, in terms of quantity. In addition, it is easy to operate small tanker trucks through narrow roads in the more congested neighborhoods.

Water Extraction and Treatment Mechanisms

Water sources in the extraction locations are mostly groundwater extracted either from shallow or deep aquifers. Only a few locations have surface water sources. At locations, such as Chobar and Baikhu, water extractors fetch water from the local stream using temporary stone intakes. In other locations, such as Jorpati, Manamaiju VDC, which are close to rivers, the extractors use shallow borings with the depth ranging from 30 feet to 150 feet. At places, such as Balaju and the Gongabu Bus Park, there are deep boring installations with a depth of more than 200 m. In the Matatirtha VDC, water extractors use dug wells as their source, the depths of which range from 20 feet to 50 feet.

Water extractors use simple water treatment facilities at different locations. The treatment usually includes aeration, sedimentation and filters (generally pressurized sand filters). Some extractors use bleaching powders and *fitkiri*[2] for chlorination. In places, such as Matatirtha VDC and Baikhu, water is pumped directly into the tanker trucks for delivery. When questioned about the need for water treatment, the well owners at the Matatirtha VDC stated that the water was of good quality and did not require any kind of treatment.

Volume of Water Extraction

There is an uncertainty regarding the exact number of tanker trucks operating in Kathmandu for the delivery of water during

Table 15.1: Water production for tanker supply in various locations inside the Kathmandu Valley

Water Extraction Locations	Number of wells	Average number of trips per well per day	Average carrying capacity of tankers per trip	Water production (liters/day)	Water production (mld)	Remark*
Peak Season						
Chobar	4	70	9,660	27,04,800	2.7	61% L and 39% S
Matatirtha VDC	12	14.5	6,000	10,44,000	1.0	100% S
Jorpati	9	76	10,020	68,53,680	6.9	67% L and 33% S

(contd.)

Table 15.1 (contd.)

Water Extraction Locations	Number of wells	Average number of trips per well per day	Average carrying capacity of tankers per trip	Water production (liters/day)	Water production (mld)	Remark*
Balaju (Ganga Hall)	1	77.5	9,000	69,7,500	0.7	50% L and 50% S
Mana-maiju VDC	4	80	6,000	19,20,000	1.9	100% S
Gothatar	3	60	10,000	18,00,000	1.8	AV
Gongabu Bus-park	1	60	10,000	6,00,000	0.6	AV
Total				1,56,19,980	15.6	
Off-peak season						
Chobar	4	60	9,240	2,2,17,600	2.2	54% L and 46% S
Matatirtha VDC	12	10.67	6,000	7,68,240	0.8	100% S
Jorpati	9	36	9,840	31,88,160	3.2	64% L and 36% S
Balaju (Ganga Hall)	1	25	9,000	2,25,000	0.2	50% L and 50% S
Mana-maiju VDC	4	25	6,000	6,00,000	0.6	100% S
Gothatar	3	25	10,000	7,50,000	0.8	AV
Gongabu Bus-park	1	25	10,000	2,50,000	0.3	AV
Total				79,99,000	8.0	

Source: Field visits and interviews 2009.

Note: Well refers to single well or boring or a group of boring. * = Weightage given to large and small tankers on the basis on number of trips they make from the source

L= Large tanker trucks of 12,000 liters and S= Small tanker trucks of 6,000 liters, AV=Average tanker capacity=10,000 liters.

peak season (mid-January to mid-June), when the water scarcity is maximum, and off-peak (rest of the year) seasons. Functionaries of the tanker entrepreneurs' association estimate that 400–450 tankers (425 on average) operate in Kathmandu, delivering an average of 10,000 liters of water per trip. Besides, in a day, a tanker makes five to seven trips (six on an average) in the dry season and one to three (two on an average) trips in the off-peak season—estimating to about 25.5 million liters per day (mld) of water transported by the tanker trucks during the dry season and 8.5 mld otherwise. A significant amount of this water (about 61 percent), during the dry season, is transported from locations such as Chobar, Matatirtha VDC, Jorpati, Balaju, Manamaiju VDC, Gothatar, and the Gongabu Bus Park. An assessment of water extraction from these locations during peak and off-peak seasons is given in Table 15.1. The remaining 39 percent of water (during the dry season) is supplied from other locations, scattered in the urban and rural areas.

The estimated water demand in Kathmandu is 280 mld. The official dry season supply by the KUKL (in 2009) was 105 mld, which will come to about 65.1 mld if we account for the 38 percent losses (including conveyance and distribution) in the system (KUKL 2009a). By comparing this amount of supply by KUKL, the water supply by utility, with the water supply of 25.5 mld by private tankers (private water supply), it is equivalent to 39 percent of the water supply of the KUKL. (Note: This is just comparison of water supply by the utility and private water supply by tanker trucks). During 2000–1, private tanker water supply was estimated equivalent to be 19 percent of the estimated 80 mld (of the NWSC) including 60 percent losses of 80 mld (Moench 2001). When we compare these two figures, we find that the contribution of tanker supply has steadily increased from 19 percent (during 2000–01) to 39 percent (in 2009), over a period of nine years. In the dry season, the tanker supply is estimated to fulfil 9.1 percent of the total estimated demand of 280 mld in Kathmandu, while the KUKL supply fulfils 23.25 percent of the demand. This shows the significant contribution of tanker supply in trying to meet the total water demand in Kathmandu.

Price of Water

The price of water, charged by tanker operators to consumers, is based on the size of the tanker, type of the source and the distance of the delivery point from the source–(*a*) The size of the tanker controls the cost incurred by tanker operators in loading of water, paying for the personnel engaged and in the repair and maintenance of trucks; (*b*) Nature of the water source affects the loading charge, for instance: the loading charge for a small truck from deep boring (in Balaju) is NRS 200 (US$ 2.86) where as it is NRS 150 (US$ 2.14) from natural spring and dug-wells at the Matatirtha VDC; and (*c*) Distance from the source to the consumer determines the driver's allowance per trip, fuel consumption and the frequency of repair and maintenance of the tankers. In general, water from a single small tanker (6,000 liters) is sold at NRS 1,200 (US$ 17.14) and water from a single large tanker (12,000 liters) at about NRS 2,000 (US$ 28.6). This makes the price of water NRS 0.2 per liter (US$ 2.86 per 1000 liters for a small tanker) and NRS 0.167 per liter (US$ 2.39 per 1000 liters for a large tanker), that is NRS 0.183 per liter (US$ 2.61 per 1000 liters) on an average. The KUKL charges NRS 17.50 (US$ 0.25) for 1,000 liters of water supply from a ½-inch pipe (KUKL 2009b). The KUKL has fixed the tariff according to diameter of the pipeline connected to household. This implies that the price of water by a private tanker is about 10.5 times higher than KUKL's piped water supply.

The value of financial transactions associated with an estimated 25.5 mld of tanker water supply in the dry season (for 150 days of water supply over five months) is about NRS 700 million (US$ 10 million), at an average rate of NRS 0.183 per liter (US$ 2.61 per 1000 liters). Similarly, in off-peak season, the value of financial transactions estimated at 8 mld of supply is about NRS 315 million (US$ 4.5 million). Thus, the total annual financial transactions associated with the tanker water market, inclusive of the volume of supply in the peak and off-peak seasons, is approximately NRS 1,015 million (US$ 14.5 million), which is approximately 4.5 times higher than the annual expenditure of the KUKL for the fiscal year 2007–08, which was NRS 225,964,157.54 (US$ 3.23 million) (KUKL 2009c).

Water Distribution Points by Tankers

The points of water distribution by the tanker trucks (see Figure 15.1) are generally concentrated in the more urbanized areas within the Ring Road though the tanker operators also supply water to institutions and households elsewhere. The locations which face serious water shortages during the dry season and depend on tanker supplies include Sundhara, Baghbazar, Newroad, Kalimati, Bafal, Teku, Lalitpur, Nayabazar, Sanepa, Dhobighat, Baneshwor, Koteshwor, Kuleshwor, Airport, and Thamel. With the expansion of urban areas outside the Ring Road and an increasing population concentration at these locations, tanker supplies have also expanded to locations, such as Chabahil, Boudha, the Pashupati area, Kalanki, Nakhipot, Balkhu, Naikap, Balaju, Tinkune, and Gongbu Bus Park, which are located outside the Ring Road. The delivery points have also expanded to distant locations including Nagarkot (more than 20 km from Koteshwor Ring Road) during the dry season. Even though consumers of the water supplies are private residents, a significant volume of the supplies goes to catering to the needs in the hotels, restaurants, schools, construction works, pharmaceutical industries, and soft drink manufacturers. Hotels and restaurants are the largest consumers of tanker water supplies.

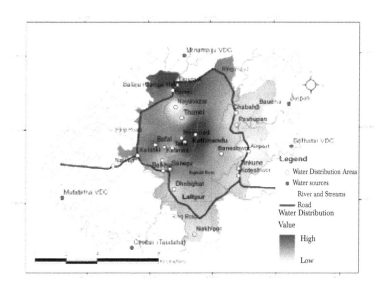

Map 15.1: Water sources and spatial variation of water distribution by tankers

Source: Field study 2009.

Quality of Water from Tanker Water Supplies

As stated earlier in the chapter, water extractors were using some form of limited treatment practices at a few sources, though the practice of chlorination of water was less frequent. The tanker water supply seems more focussed on meeting the water demand in quantity than ensuring the quality because the bulk of the demand comes during the dry season when the piped water supply is found wanting. To ascertain the quality of water delivered by the tankers from different sources, tests were carried out on samples collected from five different locations, obtained before loading the water to the tanker trucks or just after the loading of water. The results and their comparison with the National Drinking Water Quality Standard (NDWQS), 2005, are given in Table 15.2. The NDWQS 2005 was adopted by the Government of Nepal to regulate the quality of water supplied by the KUKL as well as by private sectors (such as packaged drinking water and tanker water supply).

The test results indicate that the groundwater pumped from shallow boring at Jorpati contains ammonia and iron beyond the permissible limits. Similarly, groundwater from a dug well at Matatirtha and surface water from Chobar contains iron beyond the permissible limits. Concentration of arsenic in water from all the sources was either very low or not detected in any case. Besides, all sources, except the deep boring at Balaju, showed contamination with coliforms and *E.Coli*. Presence of *E.Coli* at the source directly relates to the possible health consequences to the consumers depending on the tanker supply, considering that the existing practice of water treatment by the water entrepreneurs is unreliable.

There is also possibility of contamination of water inside tanker trucks during transportation and delivery with the residual water inside tanker trucks contaminating the fresh load of water. However, the analysis of water quality at the delivery point of the tanker trucks was not conducted because of the difficulty in tracking consumers—due to its large number and wide dispersion in Kathmandu—and the restriction in time and budget for the study.

Regulation of Water Tanker Operation in Kathmandu

Water supply by private tanker suppliers is playing a significant role in the water supply system of Kathmandu. However, the

Table 15.2: Water quality analysis of water from different sources

Sources and Samples

Parameters	Unit	Jorpati		Matatirtha		Bal-aju		Bai-khu	Cho-bar	NDWQS*
Location		Jorpati		Matatirtha		Bal-aju		Bai-khu	Cho-bar	
Source		Shallow Boring		Dug wells		Deep Bor-ing		Spr-ing	Natural Str-eam	
Date		Nov 11–14		Nov 24–27		Dec 17–19		Dec 17–19	Dec 23–26	
Physico-chemical analysis										
pH (19° C)	–	6.76	6.59	8.13	7.3	7.66	7.58	8.18	8.19	6.5–8.5
Turbidity	NTU	3.27	3.27	<1	2.18	<1	1.31	<1	2.62	5 (10)
Alkalinity as $CaCO_3$	mg/L	110	50	128	88	132	160	152	110	–
Total Hardness as $CaCO_3$	mg/L	100	50	140	96	144	164	170	122	500
Chloride	mg/L	18	6	2	3	2	1	1	2	250
Ammonia	mg/L	4.29	1.02	0.11	0.12	0.13	0.57	0.16	0.18	1.5
Nitrate	mg/L	ND	2.14	2.59	4.83	2.71	0.69	5.15	1.4	50
Iron (Fe)	mg/L	0.47	0.23	0.18	1.08	0.17	0.11	0.06	0.44	0.3 (3)
Manganese (Mn)	mg/L	0.32	0.21	ND	ND	ND	ND	ND	ND	0.2
Arsenic (As)	mg/L	0.005	ND*	ND	0.005	ND	ND	ND	ND	0.05
Microbiological Analysis										
Total Coliform	CFU/100ml	TN TC*	428	187	TN TC	111	0	250	5800	0
E. Coli	CFU/100ml	0	3	166	34	0	0	150	1100	0

Source: Field study 2009.

Note: *NDWQS = Nepal Drinking Water Quality Standard (2005), TNTC= Too numerous to count. ND= Not detected.
Values in NDWQS is the maximum permissible limits.

government has not yet acknowledged its role formally; therefore, there is no regulation of the price or the water quality. According to Moench and Janakarajan (2006), Kathmandu's water tanker operation and the market are "informal," fragmented and unregulated. Under the Water Supply Management Board Act, 2006 (2063 BS[3]) the responsibility of regulating the water services in the Kathmandu Valley, including the private water market and water extraction at the commercial scale, has been handed to the Kathmandu Valley Water Supply Management Board (KVWSMB). Under the board, the processes and procedures to regulate the water extraction and tanker operation are being implemented. To regulate the water tanker operation, private water operators formed their own association–Kathmandu Valley Drinking Water Tanker Entrepreneurs Association (registered with the Department of Cottage and Small Industries under Ministry of Industry)–in 2000 as a representative organization of water tanker entrepreneurs to develop co-operation and networking among tanker operators and to stand united in case of any insecurity and restrictions imposed by the government or other parties. One of the objectives of this association is ensuring quality services to the consumers; however, it has been more concerned with expanding the business and monopolizing the price structure.

There are at present 116 water tanker entrepreneurs with 217 tankers registered with the association. This number does not represent the actual number of entrepreneurs and tankers. On the contrary, the association estimates about 400 to 450 tankers operating in Kathmandu at present, delivering water from different locations. The association has listed the tanker entrepreneurs based on the locations of their business, such as those in Balaju, Godavari, Matatirtha, Swayambhu, Chovar areas, and Jorpati area (see Table 15.3).

The executive committee of the association has 13 functionaries that include the president, vice-president, secretary, joint-secretary, treasurer, joint-treasurer, and seven members. Entrepreneurs, who are members of the association, must abide by its constitution. To obtain the membership, entrepreneurs have to submit the name, address, phone number, and legal documents of the ownership of the tanker trucks and the sources for the water extraction. The entry fee to obtain the membership is NRS 20,000 (US$ 285.7) for the new entrepreneurs and NRS 10,000 (US$ 142. 8) for

Table 15.3: Number of water tanker entrepreneurs
and tanker trucks

S. No Regular	Area	Number of Water Tanker Entrepreneurs	Number of Water Tankers
1.	Balaju	20	30
2.	Godavari	8	19
3.	Matatirtha	8	14
4.	Swayambhu	9	16
5.	Chobar	32	56
6.	Jorpati	27	69
Irregular*		12	13
	Total	116	217

Source: Field visit, 2009 (based on records of Valley Drinking Water Tanker Entrepreneurs Association's records as of October, 2009).
Note: Irregular: Either entrepreneurs on this category have closed their business or they have not been paying the monthly fees to the association.

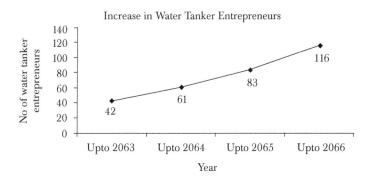

Figure 15.1: Trend of increase in number of water
tanker entrepreneurs

Source: Field visit, 2009 (based on records of Valley Drinking Water Tanker Entrepreneurs Association's records as of October 2009).
Note: 2063 BS–2006–07 CE, 2064 BS–2007–08 CE, 2065 BS–2008–09 CE, 2066 BS–2009–10 CE.

the old entrepreneurs. In addition, the association charges the membership renewal fee of NRS 1,000 (US$ 14.29) annually. The association does not fix the price of water; rather the tanker operators themselves fix the price of water transported from different locations. There has been a steady increase in the number of water tanker entrepreneurs in Kathmandu since the 1990s. The trend of the increase, over the past four years (from 2006–07 to 2009–10), based on the information provided by the association, is shown in Figure 15.2.

Along with efforts of the KVWSMB and the tankers' association, the local authorities (in different locations) have initiated processes in regulating the commercial water extraction and water tanker operation. In accordance with the Local Self Governance Act, 1999, that empowers local bodies to regulate the water business of any form within their jurisdiction, the Matatirtha VDC has formulated such an agreement–that is, the local body, market entrepreneurs and people have come together (at consensus) to carefully utilize, conserve and manage the local water resource. According to the agreement, water entrepreneurs have to pay a tax and there is a prohibition in deep boring water extraction.

However, despite the efforts to regulate the water business in Kathmandu, the presence of water entrepreneurs is hardly felt. Almost all the well owners and tanker operators, when asked about their acquaintance with the rules and regulations, knew about the registration of the tankers in the Department of Cottage and Small Industries under the Ministry of Industry and also talked about the necessity of permission from the VDCs to operate the business. However, they were not aware of the Water Supply Management Board Act, 2006, and other laws. Only the officials of the tanker association were aware of such regulations and they accepted that the government does not enforce laws. Government bodies also focused on the presence of acts and laws but failed to give evidence of its effectiveness on implementation, especially the tanker business. Nevertheless, water entrepreneurs claimed that there should be provisions to get a license to formalize their water business.

About the water quality concern, the well owners claim to test the water periodically and privately. They do not show any knowledge of monitoring of water quality from the government. The tanker association also does not have means for quality control.

Conclusion

This chapter shows that private water vendors are supplying reliable (in terms of prompt service) and significant amount of water to meet the demand of Kathmandu but at a higher price than that provided by the water supply board. However, the quality of water also needs to be considered along with the quantity delivered to the consumers. The lack of formal recognition of the existing water market by the government has restricted monitoring and evaluation of water quality that allows tankers to load water directly from the natural streams without adequate or no treatment. This continued and unregulated water extraction by the water entrepreneurs can have serious implications for the sustainability of the water sources in the valley and needs further investigation.

The water market, as a whole, is still functioning beyond the jurisdiction of government agencies in Kathmandu despite the existence of the regulatory instruments present to control private water markets. However, examples set at the local level, such as the establishment of the Matatirtha VDC, can be replicated in other locations. Nonetheless, implementation of regulatory mechanism remains uncertain as the bureaucracy fails to recognize the growing role of the water market in meeting the water needs of the urban population in Kathmandu.

Water extraction in the rural areas has already started to show its impacts on the people. Issues regarding the ownership of the resources and prioritization of water use have emerged in these locations. For instance, in Godavari area, increased water tanker business ultimately incited the people to raise their voice against such business leading to the prohibition of the extraction of water from the area. Similarly, in the Matatirtha VDC, regulatory instruments (through agreements) have been enforced to regulate the water business. Dynamics of such action and intervention are likely to be an important reality in the future and constitute a subject for future research.

Acknowledgements

The authors are grateful to Ms Shovana Maharjan for her continuous support in carrying out this study. The authors would also like to acknowledge Mr Robert Dongol, Ms Shreya Bajimaya, Mr Rikesh

Shrestha, and Mr Yvan Marcos Lopez Gonzales for their support. Finally, the authors would like to thank the Center of Research for Environment, Energy and Water (CREEW), Nepal, for providing the grant to carry out this study.

Notes

1. 1 US$ is equivalent to NRS 70.
2. Nepali (or local) word for alum, a white water treatment chemical.
3. BS is *Bikram Sambat*–the calendar system officially followed in Nepal.

References

Central Bureau of Statistics (CBS). 2003. *Population Monograph of Nepal 2003*, Vol. I. Kathmandu: CBS. Available at http://www.cbs.gov.np/ Population/Monograph/Chapter%2010%20%20Urbanization%20 and%20Development.pdf (accessed October 20, 2009).

Gyawali, D. 1988. *Preliminary Economic Analysis of a Water Loss Control Program For Kathmandu/Patan System*. Kathmandu: East Consults.

Kathmandu Upatyaka Khanepani Limited (KUKL). 2009a. "Presentation Slides: Namaste and Welcome to Kathmandu Upatyaka Khanepani Limited (KUKL)." Kathmandu: KUKL.

———. 2009b. "Notice: 2066 Poush 1 Gate Dhekhi Lagu Bhayeko Mahasul Dar." Kathmandu: KUKL. Available at http://www.kathmanduwater. org/notice/New%20Tariff%20Rates.xls/ (accessed March 13, 2010).

———. 2009c. *Annual Report on Condition and Operation of the Service System for FY 2064/65 (2007/2008)*. Kathmandu: KUKL. Available at http://www. kathmanduwater.org/reports/Annual%20Report%20%28Condition%20 &%20Operations%29%28FY64_65%29.pdf (accessed March 13, 2010).

Moench, M. and S. Janakarajan. 2006. "Water Markets, Commodity Chains and the Value of Water," *Water Nepal*, 12(1): 114–81.

Moench, Y. 2001. *Water Dynamics in a Three Part System: Investigation of the Municipal, Private and Traditional Public Water Supply Systems in Kathmandu, Nepal*. Kathmandu: Institute of Social and Environmental Transition and Nepal Water Conservation Foundation.

Pradhan, R. 2003. "A History of Water Management in Nepal: Culture, Political Economy and Water Rights," in R. Pradhan (ed.), *Law, History and Culture of Water in Nepal*, pp. 17–62. Kathmandu: Legal Research and Development Forum (FREEDEAL).

Vajracharya, D. 1973. *Licchavi Kala Ka Abhilekha (Inscriptions of the Licchavi Period)*. Kirtipur: Institute of Nepal and Asian Studies.

16

Evaluation of Institutional Arrangements for Governance of Rivers Surrounding Dhaka City

M. Shahjahan Mondal,
Mashfiqus Salehin and Hamidul Huq

——

Dhaka City, the capital of Bangladesh, is a megacity with a current population of about 15 million. It has been the principal urban center in Bangladesh (East Pakistan or East Bengal) since the 7th century CE. It became a municipality in 1864 and a corporation in 1978. The location of Dhaka in the center of Bangladesh, its historical prevalence as the administrative center of the region, the migration of rural people to Dhaka for jobs (particularly in the garment sector), growth of commerce and industries along with the population growth, and the recent real estate boom are among the important factors that have led to its growth. The adoption of embedded liberal policies since Bangladesh's independence in 1971 and neoliberal policies later on, through a number of business and investment promotional agencies including Bangladesh Export Processing Zone Authority, has further contributed to the growth of the City (Hossain 2008). The city has expanded mainly towards the north. It is surrounded by the rivers Buriganga, Turag, Balu,

Sitalakhya, and Tongi Khal (see Map 16.1), which make a circular waterway around the city and are the lifeline for its sustenance. They are an important source of water for the city-dwellers, provide navigation routes for cheap transportation of goods and passengers, and serve as recharge areas for the city groundwater aquifer and

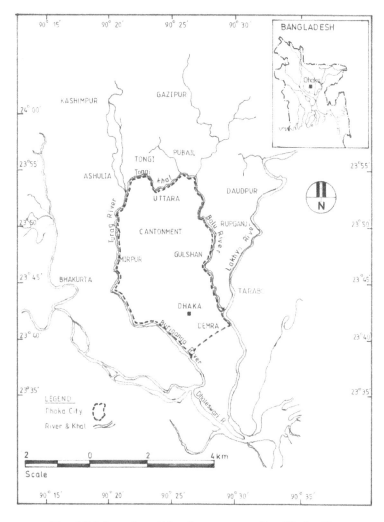

Map 16.1: Location of the rivers surrounding Dhaka City

Source: Prepared by the authors.

as natural assimilator of pollution and carriers of drainage water. However, due to physical encroachment on the rivers by land developers, lease holders, private individuals, government agencies (and the like), the rivers are gradually shrinking in width, as well as constantly are being polluted by the outflow of sewage from the well-established industrial clusters. The main polluters are industries such as tanneries, textiles, dyeing, printing, washing, and pharmaceuticals. Increasing filth and human waste have turned the water into a black gel; it is so polluted that people do not wish to touch it. Rowing across some stretches is also difficult because of the terrible stink.

There are a number of policies, acts and rules related directly or indirectly to the governance of the rivers surrounding Dhaka. Some of these are for pollution control, some for regulation of development activities and others for control of river encroachments. A number of organizations were established, entrusted with the responsibility to manage the rivers through controlling and mitigating environmental pollution, restricting river encroachment and unwanted development activities in the designated flood flow areas, and controlling urbanization. However, it appears that development activities are still continuing, the rivers are contiguously shrinking and the quality of their water is deteriorating. In the past, the government constituted a number of task forces and committees to suggest measures to tackle this pollution and encroachment of the rivers. The task force with the Minister of Shipping as its Chair provided its recommendations in 2003 in four categories: (*a*) river bank management and protection, (*b*) illegal occupancy, (*c*) pollution, and (*d*) maintenance of flow and navigability. A committee with the secretary of the Ministry of Environment and Forest (MoEF) as its Chair provided its recommendations (short, medium and long-term) in December 2002. Another committee, formed in March 2008, made 50 recommendations regarding preventive and curative measures and integrated action plan for the rivers around Dhaka City (RPMC 2008). A task force of the Ministry of Water Resources (MoWR) formulated a number of recommendations for protection of the navigation, natural course and flow of the Buriganga and to create public awareness on this issue (MoWR 2008).

While a review of these reports reveals that the sources of pollution and causes of encroachment studies have been well-captured and necessary recommendations been made, there has been very slow implementation (or in some cases, not at all) of the recommendations due to lack of enforcement, complex social and political issues as well as litigation and court cases. The principal reason is that the nature of the problem is related to governance aspects while the recommendations were mainly technical. Consequently, a study was conducted with the specific objectives of (*a*) identifying the strengths and weaknesses in the present governance regime of the rivers surrounding Dhaka city; and (*b*) recommending an appropriate governance framework for the rivers considering the hydrology, ecology, environment and socio-political situation of the region.

Research Methodology

The methodology of the study was based on a review of the existing literature, discussions with relevant public line agencies, and non-governmental and civil society organizations. The full details of the study can be found in Mondal et al. (2010). The study principally focused on the governance aspects of the five rivers surrounding Dhaka and evaluated the current institutional set-up in light of the characteristics of good governance–that is, participation, accountability, equity, effectiveness, and coherence (Rogers and Hall 2003). Though capability (resource, knowledge and skill) is not regarded as one of the characteristics, it is a prerequisite and was considered in this study.

Relevant data and information were gathered through discussions with and interviews of the relevant government and non-government officials, workers and other professionals during the period 2008–10. The organizations visited include the Department of Environment (DoE), Bangladesh Inland Water Transport Authority (BIWTA), Dhaka City Corporation (DCC), Capital Development Authority (RAJUK), Bangladesh Water Development Board (BWDB), Directorate of Land Records and Survey (DLRS), Environment Court, Deputy Commissioner's Office at Narayanganj, Survey of Bangladesh, the MoEF, Ministry of Information, International Training Network, Bangladesh Environmental Lawyers Association, Bangladesh Environment Movement, Save the Environment Movement, Bangladesh Center for Advanced Studies, as well as a few academic institutions. The views of individual industries regarding the relocation and pollution issues were also gathered.

Seven visits were made to the river sites to see the pollution and encroachment problem physically.

Secondary data on water quality, industrial establishments, types and extent of physical encroachments, uses and users of rivers (among others) were collected from relevant organizations and reviewed. The feasibility of formation of a new river commission/ body, delegation of more power and resources to the regulatory agencies (such as the DoE and BIWTA), and the possibility of environmental policing, enforcement of economic principle (polluter pays) and innovative regulations (such as pollution permit trading) were also considered for an assessment of their effectiveness. The enforcement of economic principle has the advantages of cost effectiveness, and dynamic and allocative efficiency (Leek and Lohman 1996). The concept of water authority is not new and such authorities exist in India, Sri Lanka, Vietnam, the USA, the UK, Ghana, and many other countries (see IWMI 2006). Different institutional arrangements including polycentric governance, public participation and bioregional perspective were also considered (Huitema et al. 2009). The role of the judiciary and political and institutional leadership in protecting the rivers from deterioration were evaluated. Finally, the current water flow condition and its trend were studied.

River System in and around Dhaka

The Dhaka watershed comprises an area of about 1700 km² (RPMC 2008). The river system in and around the city includes the Buriganga, Turag, Tongi Khal, Balu, Sitalakhya, and Dhaleshwari. Some of the salient features of the river system are presented in Table 16.1. The total length of the rivers surrounding Dhaka and Narayanganj is about 110 km (ibid.). All the rivers are tidal and perennial. The Buriganga offtakes from the Turag at Mohammadpur and falls into the Dhaleshwari at Fatullah, Narayanganj (BWDB 2005). The Turag originates from the terrace areas of Kaliakoir. The Tongi Khal offtakes from the Turag and empties into the Balu; the latter offtakes from the Tongi Khal at Tongi and falls into the Sitalakhya at Demra. The Sitalakhya originates from the confluence of the Old Brahmaputra and Banar at Shibpur, Narsingdi, and falls into the Dhaleshwari. These rivers together form a circular water way around the city.

Table 16.1: Salient features of river system in and around
Dhaka City

Rivers	Length (km)	Average width (m)	Catchment area (km^2)	Discharge (m^3/s)
Buriganga	45	265	253	50–150
Turag	71	218	1021	124–1136
Tongi Khal	17	60	35	35–205
Balu	45	100	722	60–744
Sitalakhya	73	273	3803	195–2742

Source: BWDB 2005.

Present State of the Rivers

Water Pollution

The Dhaka watershed contains nearly 7,000 industries and has nine clusters of industrial pollution–Tongi, Hazaribagh, Tejgaon, Tarabo, Narayanganj, Savar, Dhaka Export Processing Zone (DEPZ), Gazipur, and Ghorashal. Table 16.2 provides an indication of the pollutant load generated from each cluster. It is seen that about 1.5 Mm^3 of effluents are generated in these industrial clusters and the Biochemical Oxygen Demand (BOD) for them is about 0.4 Mkg of Oxygen.

The Buriganga is the main outlet of all types of wastes generated in Dhaka as most of the untreated sewage is discharged here. A significant source of its pollution is the waste emission from more than 200 tanneries in the Hazaribagh area (categorized as "red"–that is, the most dangerous), which do not have any effluent treatment plants (ETPs). As seen in Table 16. 3, the BOD load in the Buriganga is 45 mg/l, much higher than the maximum allowable limit of 3 mg/l in surface water for conventional treatment for drinking water supply (MoEF 1997). The Dissolved Oxygen (DO) level was found to be below 1 mg/l, whereas the DoE recommends the level to be above 6 mg/l. It has to be pointed out here the DO level below 2 mg/l causes the death of fish and other aquatic lives. This is the reason why the Buriganga is now called a biologically dead river. The presence of heavy metals can also be seen in the table. However, the conventional treatment system in Bangladesh (for water supply) is not designed to treat heavy metals.

Table 16.2: Biochemical oxygen demand (BOD) into the
Dhaka rivers from selected industrial clusters

Cluster	Volume of Effluent (m³/d)			BOD (kg/d)				
	Indus-trial	Dom-estic	Total	Indus-trial	Dom-estic	Total	Dom-estic ret-ained	Total gener ated
Tongi	21708	13450	35158	3797	3362	7159	5396	12555
Hazari-bagh	49489	37695	87184	46349	9424	55773	10891	66664
Tejgaon	157853	71280	229133	41791	17820	59611	11364	70975
Tarabo	84672	0	84672	26962	0	26962	17854	44816
Naray-anganj	456225	38721	494946	33344	9681	43025	31932	74957
Savar	7738	1376	9114	1413	344	1757	6534	8291
Gazipur	192845	0	192845	18922	0	18922	1043	19965
DEPZ	314755	0	314755	31042	0	31042	17071	48113
Ghor-ashal	44928	0	44928	5422	0	5422	10428	15850
Total	1330213	162522	1492735	209042	40631	249673	112513	362186

Source: Institute of Water Modelling 2006.

The Sitalakhya receives the discharge of the Balu, untreated
wastes of industries from Kanchan to Narayanganj and the DND
area, and municipal and human wastes from Narayanganj, which
are discharged through six drains–Majhipara, Killerpool, Kalibazar,
Tanbazar, BK road, and the DND *khals* (open drainage channels).
The DO value was recorded to be 2.9 mg/l at the intake point of the
river in March 2003 and 2 mg/l near Narayanganj in March 2005.
However, its water quality is relatively better than the three rivers
mentioned earlier. The water of the Dhaleswari is not seriously
contaminated either and it has a relatively higher DO level and
lower BOD and other parameters (see Table 16.3)

Encroachment of Rivers

Encroachment narrows the river width, reduces the water flow,
and disturbs the river morphology, flora and fauna. Such reduction
in water flow also decreases the pollution assimilation and dilution
capacity of the rivers. Over the years, the rivers around Dhaka have

Table 16.3: Water quality of the rivers surrounding Dhaka City
during the lean period

Parameter	Buriganga	Turag	Balu	Sitalakhya	Dhaleswari
Acidity (pH)	6.99	6.94	7.04	7.27	6.81
Ammonia-nitrogen	0.05	0.01	0.03	0.01	0.00
Ammonium-nitrogen	9.90	1.17	4.93	0.56	0.29
Total Ammonia	9.95	1.18	4.96	0.57	0.29
Nitrate-nitrogen	3.0	0.4	0.3	1.1	2.4
Total dissolved solids	608	251	456	130	285
Total suspended solids	13	51	52	36	23
Phosphate	4.63	0.51	1.37	0.36	1.47
Sulphate	115	47.2	74.5	16	36.6
Biochemical Oxygen Demand	45	25	30	5	13
Chemical Oxygen Demand	73	41	51	10	22
Aluminium	0.156	0.156	0.159	0.111	0.134
Cadmium	0.021	0.003	0.023	0.023	0.001
Lead	<1	16.2	14.3	<1	<1
Chromium	116	30	82	11	58

Source: IWM 2006.
Note: all parameters are in mg/l except for pH.

been subject to indiscriminate encroachment by land grabbers at more than 3,000 locations (RPMC 2008). The BIWTA evicted 2,862 encroachments from the river banks up to the year 2008; yet, there are more than 400 encroachments. The occupiers usually belong to the influential section of society and use their political and monetary power to manage the government machineries. Their network has infiltrated deep into the governance system and a section of the public institutions facilitate such infringements by preparing fake documents, without taking any action. Therefore, it is a challenge to tackle the situation. The BIWTA sometimes carries out eviction drives with the help of district administration and police but meets with little success due to the lack of protection and management of the recovered lands.

Water Flow in Rivers

A number of studies reported that the discharge of the rivers is reducing gradually due to the closure of the spill channels of the

Jamuna River (internationally, the Brahmaputra River) because of siltation. Furthermore, sediments from different sources–such as dredged spoils, land erosion, construction site, and land development–enter the canal and river system causing reduction of the normal depth, storage and conveyance capacity of the rivers. As a result, pollution from different sources is not flushed out from the river system and remains stagnant. Sedimentation is, thus, one of the major reasons for increasing pollution levels in the rivers, especially during the lean season (November–May). In the past, the BIWTA had initiated a project that excavated the rivers (in the waterway), built launch terminals at eight locations, and initiated cargo and passenger services. But the project was not very successful mainly due to the inadequate depth of the navigation route to ply the vessels. Currently, the Authority is also implementing a river-cleaning project without much success due to gaps in planning and organizational collaboration and co-ordination.

Impact of Poor River Condition

The alarming condition of the rivers around Dhaka has put natural resources and the social lives of the people living around the riverside in grave danger. Due to the nuisance caused by the physical environment, people are forced to migrate to other places or remain isolated from their relatives and friends. It also affects their social relations–as people from other parts of the country show less interest in building marital ties with the riverside people (RPMC 2008). The Buriganga, once considered as the pride of Dhaka and admired for its aesthetic view, has now lost its beauty and attraction and is no longer a place of recreation.

More than 3–4 million people suffer the consequences of poor water quality in the river system caused by various types of industrial untreated wastes (RPMC 2008). There is a correlation between the incidences of skin diseases, diarrhoea and dysentery and the level of river pollution. About 80 percent of the population suffers from these diseases compared with the 30 percent or lesser in a control area. Consequently, the treatment cost is about 2.5 times in the affected area (World Bank 2008). Some households suffer considerable loss of income when they are unable to work due to sickness.

About 75 percent of Dhaka is under the water supply network of Dhaka Water Supply and Sewerage Authority (DWASA). The source of water is 83 percent from groundwater and the rest is from surface water. There are about 485 deep tube wells, two major treatment plants–at Saidabad and Chadnighat, respectively–and two smaller treatment units at Narayanganj. Water is so polluted at the intake points of the treatment plants that it cannot be treated at the drinking water level, which has led to the idea of relocating the intake points. But the financial implications would eventually be on the users. Evidence of groundwater contamination (high values of electrical conductivity) is reported in areas near the Buriganga (IWM 2006). Chemicals carried by industrial waste, such as cadmium, chromium and mercury, are creeping into the groundwater, posing a serious threat to public health.

The river pollution was found to have an adverse impact on soil and agriculture. Heavy metals, such as copper, iron, manganese, zinc, lead, cadmium, nickel, chromium and arsenic, were found to be accumulating in soils due to irrigation with polluted river water (Rahman and Mondal 2009). This, in turn, has negatively affected the growth and yield parameters of rice (for example: plant height, root length, panicle length, number of tillers per plant, number of grains per panicle, and grain yield). Furthermore, the quality of the rice in terms of colour, size and taste was also affected, which had a financial implication for the farmers; the farm income was 50 percent lower at the polluted site than that of the control site. Besides, the continuous consumption of the poor quality, low nutritional rice has been a potential risk to human health. The impact of water pollution on fish and aquatic life is also alarming. Both production and number of fish species have gradually declined. There is hardly any fish or aquatic life apart from zero oxygen survival type of organisms in the Buriganga. The cost linked only to the poor water quality of the rivers was estimated to be US$ 400 million (DoE 2004) treating the poor water quality of the rivers was estimated to be. The total cost was derived from different sectoral costs of water quality. About 40 percent, 22 percent, 21 percent and 17 percent of the total cost were from the health, industry, amenity, and agriculture and fishery sectors, respectively.

Present Institutional Set-up for Governance of the Rivers

Policy, Act and Rules

There are a number of legal instruments to protect and conserve the rivers. The Environment Policy, 1992, calls for control of unplanned housing and urbanization, and the preservation and development of wetlands. The Bangladesh Environment Conservation Act, 1995, and its subsequent amendments (in 2000 and 2002) provide mandates to the DoE for conservation of the environment, improvement of environmental standards, and control and mitigation of environmental pollution. The Environment Conservation Rules, 1997, make it obligatory to obtain location and environmental clearance certificates for different industrial units and projects periodically. It also sets a standard for sewage, waste and effluent discharge. The Environment Court Act, 2000, was enacted for the establishment of environment courts. One such court now exists in Dhaka to deal with environmental cases.

The National Land Use Policy, 2001, highlights the importance of agricultural lands and wetlands for crop and fish production, respectively, and calls for actions to arrest the on-going degradation and encroachment by horizontal urbanization, housing and industrial development. The National Agriculture Policy, 1999, states that the fertile agricultural land is going out of cultivation due to its use for non-agricultural purposes, such as private construction, house building and brick fields, and calls for appropriate measures to stop this trend.

Line Agency

There are a number of public line agencies to protect the rivers from pollution and encroachment and to maintain their flow regime. Setting an industrial effluent standard, providing permission for establishing new industries, and monitoring and enforcing compliance are among the principal mandates of the DoE (MoEF 1995, 1997). The BIWTA is mandated to maintain navigability of waterways, lease out river banks, and protect rivers from encroachment (Ministry of Shipping 1966, 1973). To develop the zoning map, RAJUK has been entrusted with enforcing zoning regulation and regulating infrastructural development. The DCC looks after solid waste management and drainage management while drinking water supply and waste and storm water management are under the purview of the DWASA. The DLRS has the responsibilities of maintaining land records (including *khas* lands [state-owned lands] and rivers) and to conduct land surveys as well as update land records. The offices of the deputy commissioners at Dhaka, Narayanganj and Gazipur carry out the functions of leasing out river banks and carrying out eviction drives against illegal establishments. The environment court deals with the environmental cases. Apart from these, there are other public departments whose activities are linked with the management and governance of rivers (for details, see Mondal et al. 2010).

As per the law, the Ministry of Land is the owner of the lands falling within the rivers. The MoWR is entrusted with the planning and development of water resources of these rivers as well as the protection and improvement of river banks. The Ministry of Shipping is mandated to maintain the navigability of the rivers as well as keep the rivers and banks encroachment free. The Ministry of Power, Energy and Mineral Resources undertakes sand mining and leasing activities. The MoEF works for the protection and conservation of the rivers and their ecology and biodiversity. The Ministry of Local Government, Rural Development and Co-operatives uses the rivers as effluent and waste assimilator and dumping ground.

Weaknesses in the Present Institutional Set-Up

Lack of Coherence and Common Vision

The current institutions work with their own organizational mandates and follow a segregated and sectoral approach. They do not have a shared vision for an integrated approach in environmental quality improvement, ecosystem sustenance and sustainable development. The current governance regime lacks clear policy and strategic direction and is not clear about the broad objective of development– whether it is economic development with the poor river condition, a healthy river with limited economic development, or a trade-off between ecosystem health and economic development.

The 7,000 industries located inside the Dhaka watershed contribute about 60 percent of the total pollution load to the rivers. They are, on the other hand, the biggest import sector of the country and employ a large section of the poor people. It is argued that installing and running ETPs will shoot up the production cost, making it difficult for the industries to survive in the competitive world markets. The environmentalists and the academics, however, view it as a propaganda campaign by the industrialists with the intention of making undue profit, violating social equity and environmental justice. They further emphasize that the eco-system, bio-diversity and environment should not be allowed to degrade simply for economic benefits. The polluter–pays principle would not help either.

Due to the lack of institutional cohesion and vision, the natural drains of Dhaka were allowed to be encroached, filled in and narrowed down, even by public line agencies themselves (Mondal et al. 2010). The number of drains decreased from 43 to 26 functional ones. There are some overlaps and redundancy in the institutional mandates and activities. For example, both DCC and DWASA run the drainage system of the city. The DWASA maintains 265 km pipelines, 65 km canals and 16 km box culverts, while the DCC runs 999 km open drains and 1,052 km pipe drains. The DCC maintains the primary drainage lines from households to DWASA facilities. However, there is a lack of co-ordination between the two organizations as acknowledged by both organizations (HIC 2009). Both academics and administrators, including decision-makers at the DCC and the DWASA, opined that drainage system management should be regulated by a single authority/agency. Both the DCC and RAJUK also together control the management of lakes in the City. Another example of dual responsibilities is the operation and maintenance of pumping stations–DWASA maintains four and BWDB maintains one storm water drainage stations. Similarly, the management of pumping stations and drainage and flood control sluices and gates should also be undertaken by one body.

Effectiveness

The present institutional set-up is not very effective for the maintenance of river health, functionaries and services. The DoE was established to control water and environmental pollution. Ironically, the pollution level has gradually increased. Similarly, the RAJUK was created to control and reduce unplanned development; on the contrary, these are continuing in full view of the authority and, sometimes, with its help. The DCC and DWASA are supposed to take care of the health and environment of the City but they themselves are contributing about 40 percent of the pollution load to the rivers by dumping solid wastes or discharging industrial and municipal liquid wastes. The BIWTA and the three district administrations are supposed to keep the rivers and their shores free from encroachments and obstructions. However, the rivers are being gradually encroached upon and filled in by the land developers, government agencies and individuals alike. The DLRS is supposed to maintain the ownership of the rivers and *khas* lands in favour of the government. But individuals/developers seem to produce documents claiming the lands, even inside the rivers, as their own (for details, see Mondal et al. 2010).

Accountability

Lack of accountability has also been a characteristic of the present institutional set-up. Accountability has three elements–transparency, responsiveness and compliance. The institutions are not complying with their mandates; they are neither responsive to the public concerns and suffering nor transparent about their activities and actions. For example, the information about the exact number of different categories and types of industries, their environmental compliance status, the condition of the eco-system and bio-diversity, and the list of physical encroachment and encroachers are not made public. Such disclosure of information could act as a deterrent to the illegal encroachers and polluters and might also work as an effective regulatory tool.

Most public servants and institutions (in general) are driven by self-interest and personal gain. These trends need to be arrested and these can be done by major changes in existing laws. Lessons can be drawn from the Environment Protection Act, 1986, of India (Ministry of Environment and Forests 1986), which has the provision of better accountability of government departments. Article 17 of the act gives clear directives about the actions and/ or punishments to be taken against any offence committed by any department of government. However, there is no such provision in the Environment Conservation Act and other related acts in Bangladesh to caution public line agencies and their employees, improving their effectiveness and accountability. Bangladesh passed the Right To Information Act and started its implementation on July 1, 2009. However, the information commission has not been fully functional due to lack of required office space, information delivery system, manpower, and funding.

Participation

The role of multiple stakeholders in water resources and environmental management is emphasized in international discourses (GWP 2000; Ison et al. 2007; Pahl-Wostl et al. 2007). Decision-making and implementation activities of the institutions in Dhaka watershed are not designed in a participatory framework; rather, they follow the typical top-down, command and control approach. The platform for hearing the grievances and/or voices of the users' groups, beneficiaries, polluters, encroachers, civil societies, and NGOs is limited. There is no representation of other agencies in a single line agency (Mondal et al. 2010). For example, the DWASA and DCC have no representation in the DoE and vice versa. Neither does the BIWTA comprise any representation from RAJUK and district administrations. Given the magnitude of the pollution and encroachment problems, such cross-agency representation is vital.

Capacity

There is a chronic shortage in the institutional capacity and resources of the public organizations. The DoE is understaffed to monitor compliance and enforce regulation, has limited laboratory facilities to carry out routine tests and inadequate budgetary allocations. The BIWTA is also understaffed. It has only seven, very old dredgers (whose economic lives have expired) with an annual excavation capacity of 2.6 Mm^3 against the actual requirement of 10 Mm^3. The present sewerage system, maintained by DWASA, covers 110 km^2 of 360 km^2 in the city area, serving a mere 20 percent of the population. Generation of domestic and industrial sewage has been estimated at 1.3 Mm^3 per day against the treatment capacity of 0.12 Mm^3 (RPMC 2008). The DCC has the capacity to collect only about 44 percent of the total 3.2 Mkg of generated solid waste. The shortage of conservancy vehicles, trucks and container carriers due to limited funds for utility services, management and development is believed to be one of the reasons for the poor performance of the DCC (DoE 2005). In connection to wetland management, Ishrat Islam (2008) pointed to the incapability of the institutions, responsible for protecting wetlands of Dhaka, to arrest conversion of wetlands in the restricted flood flow zones.

Towards a New Institutional Framework

The civil society organizations, non-governmental organizations, and the print and electronic media have campaigned against the pollution and encroachment of the rivers surrounding the city for a few years. Their role in bringing the issue to public notice, raising public awareness, sensitizing and mobilizing the public around the problem is commendable. They pointed out some of the institutional lapses and gaps in the current setup and even took the issue to the High Court. The apex court has also been playing an important role in the issue.

In its historic verdict in June 2009, on two writs initiated by NGOs in connection to industrial pollution, the High Court provided very specific, time-bound directives and orders to the heads of a number of public agencies to free the rivers from encroachment and pollution (The Supreme Court of Bangladesh 2009a, b). It also cautioned that the heads would be "personally accountable" if any impediment is caused in the implementation of its orders (The Supreme Court of Bangladesh 2009a: 10). The inclusion of the "accountability" clause had great significance in a later execution of its verdict. All the line agencies immediately roused into action and started planning and implementing the directives. Given the socio-political, law and order, and bureaucratic situation of the country, it would have not been possible to initiate the implementation of these court orders without the clause. This can be used as a great pointer on how to change the existing laws and acts of the country for better governance.

With the High Court verdicts and media campaigns, the present political regime is also playing a positive role in tackling the problem. In a gazette notification on August 27, 2009, the government constituted a task force with the Minister of Shipping as its Chair to provide suggestions and recommendations to maintain navigability and flow of the rivers. The task force comprises 23 members including government ministers, political leaders, legislators, lawyers, bureaucrats, public organization heads, academics, journalists, and environmentalists. It has already conducted six meetings and provided valuable directions and guidance to tackle the situation. As an inter-ministerial committee, it has the power and authority to provide direction and follow its execution. The government has declared the four rivers—Buriganga, Turag, Balu and Sitalakya—as ecologically critical areas on October 4, 2009. All these are positive changes and the people are of the opinion that the task force should be allowed to continue its work for some more time. This could ultimately pave the way for a National River Commission.

Most people believe that a River Commission, with adequate power and legal authority, can better co-ordinate the policies and activities of the different public agencies. It can work as a policy and program developer, and monitor activities and agreements. However, its focus would be more at institutional level. The current laws are mainly departmental, but the scale of the issue cuts across many sectors. The new commission can fill that gap. Nevertheless, a few government officials (who were interviewed) had reservations about the success of the commission. In their view, the network and power of the polluters and encroachers are so strong and so deeply entrenched in the political, administrative and cultural system that the solution of the problem would require a visionary and patriotic leadership, and political and institutional commitment of actual implementation of different legal instruments. Furthermore, a few academics were of the opinion that an additional delegation of authority and autonomy to the city corporation, as is the norm in developed countries like the USA, is a necessity. Members of the civil society and non-governmental organizations along with most academics and some public officials, however, felt the necessity of a formal mechanism for the co-ordination of planning and development activities and control of human behaviour in the Dhaka watershed, taking into consideration the inter-linkages and interdependence among humans, the eco-system and the environment.

Conclusions

The rivers surrounding Dhaka city are in a deplorable condition. They are highly polluted and encroached upon illegally. There are a number of policies, acts, plans, and institutions to manage the rivers. However, these are not effective enough to arrest the on-going degradation of the rivers and eco-system. The absence of good governance characteristics within the organizations is the principal reason for the poor condition of the rivers. There is a lack of coherence and vision among the organizations. They work in isolation and there is no formal mechanism of co-ordinating their activities. Constituting a National River Commission with sufficient power and authority can improve the situation by offering a broader vision, cross-sectoral integration, co-ordination platform, and accountability tools. The success of such a commission would lie with the free space given by political leadership and the legal system. Recent verdicts of the apex court, which included an accountability clause for public servants for non-compliance, and the functioning of the task force, in the spirit of the court orders and directives, can provide a general framework for the organizational structure of the commission. The different kinds of media today, civil societies and non-governmental organizations have played important roles in improving the governance of the rivers.

References

Bangladesh Environment Management Project. 2004. *Strategic Benefit-Cost Framework*. Report for the Department of Environment (DoE), Dhaka.

Bangladesh Water Development Board (BWDB). 2005. "Rivers of Bangladesh." Dhaka: BWDB.

———. 2005. "Dhaka City: State of Environment 2005." Dhaka: DoE.

Global Water Partnership (GWP). 2000. "Integrated Water Resources Management." TAC Background Paper No. 4, GWP, Stockholm.

Habitat International Coalition (HIC). 2009. "Water Logging in Dhaka after Heavy Rainfall." Dhaka: HIC. Available at www.hic-net.org/news.php?pid=3170 (accessed May 14, 2013).

Hossain, S. 2008. "Rapid Urban Growth and Poverty in Dhaka City," *Bangladesh e-Journal of Sociology*, 5(1): 1–24.

Huitema, D., E. Mostert, W. Egas, S. Moellenkamp, C. Pahl-Wostl, and R. Yalcin. 2009. "Adaptive Water Governance: Assessing the Institutional Prescriptions of Adaptive (Co-) management from a Governance Perspective and Defining a Research Agenda," *Ecology and Society*, 14(1): 26.

Islam, I. 2009. *Wetlands of Dhaka Metro Area: A Study from Social, Economic and Institutional Perspectives*. Dhaka: A. H. Development Publishing House.

Ison, R., N. Roling and D. Watson. 2007. "Challenges to Science and Society in the Sustainable Management and Use of Water: Investigating the Role of Social Learning," *Environmental Science and Policy*, 10(6): 499–511.

Institute of Water Modelling (IWM). 2006. *Resource Assessment and Monitoring of Water Supply Sources for Dhaka City*. Final Report on Resource Assessment, Vol. I: Main Report. Dhaka: IWM.

International Water Management Institute (IWMI). 2006. "Water Governance in the Mekong Region: The Need for More Informed Policy-Making." Water Policy Briefing 22. Colombo: IWMI.

Leek, F. and A. S. Lohman. 1996. "Charges in Water Quality Management in the EU," *European Environment*, 6: 33–39.

Ministry of Agriculture (MoA). 1999. *National Agriculture Policy.* Dhaka: Government of Bangladesh.

Ministry of Environment and Forests. 1986. *The Environment (Protection) Act.* New Delhi: Government of India.

Ministry of Environment and Forest (MoEF). 1992. *Environment Policy 1992 and Implementation Activities.* Dhaka: Government of Bangladesh.

———. 1995. *The Bangladesh Environment Conservation Act.* Dhaka: Government of Bangladesh.

———. 1997. *The Environment Conservation Rules.* Dhaka: Government of Bangladesh.

Ministry of Law (MoL). 2000. *The Environment Court Act.* Dhaka: Government of Bangladesh.

———. 2001. *The National Land Use Policy.* Dhaka: Government of Bangladesh.

Ministry of Shipping (MoS). 1966. *The Port Rules.* Dhaka: Government of Bangladesh.

———. 1973. *The Removal of Wreck and Obstructions in Inland Navigable Waterways Rules.* Dhaka: Government of Bangladesh.

Ministry of Water Resources (MoWR). 2008. *Report of the Task Force on Sustenance of Navigation and Restoration of Natural Course and Flow of Buriganga River.* Dhaka: Government of Bangladesh.

Mondal, M. S., M. Salehin, H. Huq, F. Azim, M. Monem, and S. A. Jahid. 2010. *Development of an Institutional Framework for Management of Rivers Surrounding Dhaka City.* Dhaka: Institute of Water and Flood Management.

Pahl-Wostl, C., M. Craps, A. Dewulf, E. Mostert, D. Tabara, and T. Taillieu. 2007. "Social Learning and Water Resources Management," *Ecology and Society*, 12(2): 5.

Rahman, M. and M. S. Mondal. 2009. "Effect of Irrigation with River Water Containing Heavy Metals on Soil and Rice Yield in Dhamrai Upazila, Dhaka," in *Proceedings of the Second International Conference on Water and Flood Management*, March 15–17, pp. 623–32. Dhaka: Institute of Water and Flood Management.

Rogers, P. and A. W. Hall. 2003. "Effective Water Governance." TEC Background Paper No. 7, Global Water Partnership, Stockholm.

River Pollution Mitigation Committee (RPMC). 2008. *Report on Mitigation of River Pollution of Buriganga and Linked Rivers: Turag, Tongi Khal, Balu, Sitalakhya and Dhaleswari.* Dhaka: RPMC.

The Supreme Court of Bangladesh. 2009a. *Special Original Jurisdiction, Writ Petition No. 891 of 1994.* Dhaka: High Court Division.

———. 2009b. *Special Original Jurisdiction, Writ Petition No. 3503/2009.* Dhaka: High Court Division.

World Bank. 2008. *Institutional and Economic Analyses of Industrial Effluent Pollution and Control in Greater Dhaka Watershed.* Final Draft. Dhaka: World Bank.

17

Sustainable Urban Water Supply and Sanitation

A Case from Kandy, Sri Lanka

Sunil Thrikawala, E. R. N. Gunawardena
and *L. H. P. Gunaratne*

—

Contamination of drinking water sources with sewerage has been one of the emerging problems in Sri Lanka. This problem has been widely reported in the central region (NRMS 2005), which uses water from the Mahaweli River as a major source. As a solution, sewerage treatment plants (STPs) are being proposed as a pre-requisite to many water supply development projects to address this problem of sewerage contamination of the source. One such project, the Kandy City Water Supply Augmentation and Environmental Improvement Project (KCWSAEIP), is supposed to provide safe sanitation by treating the effluents in the drinking water source and ensuring clean drinking water within the city. The National Water Supply and Drainage Board (NWSDB) and Kandy Municipal Council (KMC) are the two major institutions responsible for water supply and sanitation (WSS) services to Kandy city. The project will be implemented by the NWSDB and, once completed, handed over to the KMC for maintenance.

There have been many concerns about this project since its inception. Those who live close to the proposed sites of the STPs have been continuously protesting as they are genuinely anxious about the negative consequences. The loan amount, obtained for the STPs, has also been very substantial and, as a result, there have been apprehensions about the increased tariff pass on to the consumers. Though the proposed STP is of superior technology, there are serious doubts regarding the KMC's capability to operate and maintain it. Furthermore, the ability of the NWSDB in providing water supply and sanitation services with this increased debt burden, due to borrowing and resultant financial difficulties, is a major concern at the national level.

In this context, a study was conducted with the overall objective of assessing the sustainability of water supply and sanitation services in Kandy. To achieve this, it became necessary to develop a conceptual framework and methodology. This chapter describes this conceptual framework and its findings along with some of the policy directives, which will be useful in the long run.

Approach and Methodology

There has been substantial research to improve urban water management, mainly in the context of developing new technologies. However, T. H. F. Wong (2006) argues that further research on institutional issues relating to sustainable urban water management is needed. Further, S. van de Meene and R. Brown (2007) showed that even though there has been substantial investment in urban water reforms, the latter have not been as successful as anticipated– probably due to the lack of a critical analysis on the existing capacity. Concepts such as integrated water resources management (IWRM), decentralization, privatization, and demand management have been incorporated in the discourses on reform processes over the last two decades. The 5th World Water Forum (WWF), held in Istanbul in 2009, declared its stand, under six themes, on how the water crisis should be handled and in achieving the Millennium Development Goals (MDGs) (Istanbul Water Guide 2009). Under the 4th theme– Water Governance and Management–recommendations have been made on various issues, such as the right to water and sanitation for improved access, water institutions and water reforms, ethics, transparency and empowerment of stakeholders, and optimizing private and public roles in water services. The Istanbul Water Guide (2009) also discusses issues of sustainable WSS services through financing in the water sector, cost recovery strategies and pro-poor financing policies and strategies. Given this background, a schematic diagram of the conceptual framework (for the study) was developed (see Figure 17. 1).

The methodology followed in this study included data collection and analysis of the three major components given in the diagram, that is, governance, management (institutional capacity) and financing and cost recovery aspects. The study concentrated on the NWSDB and the KMC as key institutions of water and sanitation facilities to the city. Though it focused on Kandy, issues of national importance came to the fore since the NWSDB is considered as the major national institution providing WSS services to the entire country of Sri Lanka.

Major Findings

Management (Institutional Capacity)

Human Resources

Institutionally, the NWSDB has expanded its scope during the last three decades. Employee numbers have increased from 1,000 in 1975 to 8,460 in 2007 (see Table 17.1). The service connections

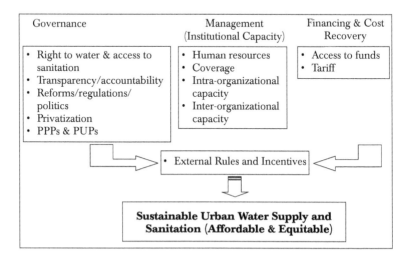

Figure 17.1: Schematics of the conceptual framework

Source: Prepared by the authors.

have also gradually increased from 0.18 million in 1989 to 1.07 million in 2007. Thus, the number of employees per connection has decreased over the years from 40 in 1990 to 7.8 in 2007, indicating that the NWSDB is expanding its capacity while increasing service efficiency.

The NWSDB has emphasized on developing the institution during the last few years. Work Improvement Teams were formed in all divisions with staff participation at all levels. It has also embarked on an extensive information technology (IT) overhaul, enabling projects to improve operational efficiency (in many core activities). It has also tried to improve its skill level through training of staff at all levels. The online payment system was continued and upgraded. Under regular activities, the department dealing with customer complaints was improved. Furthermore, a system was developed for monitoring water quality through which laboratories throughout the island will be able to forward water quality details to the head office.

In contrast, the Department of the Waterworks of the KMC undertakes all forms of production, distribution and operation and maintenance activities to meet the water demand. The office consists of 241 members which includes executive officers (waterworks engineer, assistant waterworks engineer, accountant and technical officers), skilled and non-skilled labourers. However, according to the data at the KMC, there is a shortage of 169 staff members including one assistant waterworks engineer and 95 non-skilled labourers to operate the institution at its full capacity.

The proposed Kandy STP consists of state-of-the-art technology and, hence, requires technical personnel who are conversant with

Table 17.1: Human resources development of NWSDB

Item		1990	1995	2000	2005	2006	2007
Staff (No.)	Permanent	6591	6817	7592	7661	NA	NA
	Casual	764	524	214	123	NA	NA
	Contract	0	214	43	130	NA	NA
	No pay	0	13	NA	NA	NA	NA
	Total	7355	7568	7849	7914	8350	8460
No. of Employees/ connection		40	23	13	9.2	8.8	7.8

Source: NWSDB Annual Report (1990–2007).

this technology (Narayanan and Thrikawala 2011). However, at present, there is hardly any expertise available to operate such plants locally. The KMC, which has only one engineer in the water supply department, will not be able to manage the STP unless a long-term human resources development plan is designed and implemented; this requires appointing a core group of engineers and training them to operate the STP. The funding required to recruit additional people is hard to come by since the KMC already faces difficulties in continuing their current services with limited revenues. The people at the KMC believe that they could outsource the repairs and maintenance of the STP, as and when there are problems. However, such services are not readily available in Sri Lanka as there is a dearth of technically qualified personnel in the sewerage treatment. Therefore, it is important to look at institutional arrangements since any failure of the STP could have adverse consequences.

Coverage

Over the years, the NWSDB has continued to expand its coverage of services through projects. The pipe-borne water production has been substantially increased along with the number of connections (see Table 17.2). Simultaneously, there is a decrease in the number of public stand posts, which indicates an improvement in the efficiency of the service. However, as G. T. Daigger (2009) points out, the NWSDB follows the traditional "take, make, waste" approach that could be heading towards a financially unsustainable water service of poor quality.

The NWSDB has also been broadening its services of sewerage disposal and by 2007, the sewer coverage was 2.5 percent. However, sanitation services have been confined to the capital city of Colombo and its suburbs. Apart from that, there are several small-scale STPs being operated for housing schemes and government institutions. During 2007, the NWSDB established a separate division to handle sewerage activities and appointed an Additional General Manger (AGM), who was responsible for the operations, planning and designing of foreign-funded sewerage projects. Currently, the division is handling 18 projects of various magnitudes (NWSDB Annual Report 2007). A few centralized, large-scale sewerage treatment projects have also been proposed for a few major cities in Sri Lanka including Kandy.

Table 17.2: Pipe-borne water coverage during
the period from 1990 to 2007

Item		1990	1995	2000	2005	2007
Piped water production (Million m³)		1.2	275	332	383	424
Domestic connections (000')	Greater Colombo	109.4	145.6	249.7	392.4	433.4
	Other areas	51.6	136.0	258.1	422.6	543.12
Stand Post (000')	Greater Colombo	1.5	4.58	6.6	4.4	3.3
	Other areas	1.7	4.9	5.3	3.9	3.6
Non-domestic (000')	Greater Colombo	13.4	16.1	25.5	35.8	38.9
	Other areas	6.2	16.0	35.9	48.8	56.5
Total No of Service Connections (000')		185.7	324.4	582.3	907.6	1078.9

Source: NWSDB Annual Reports (various issues), 1990–2007.

The KMC has also increased the capacity of its service to the city (see Table 17.3). The number of water connections has increased by 61 percent during the period 1995–2009. Since the relevant data on the number of employees during this period was not available, it is difficult to comment on the efficiency of the service. Currently, the KMC is unable to further increase its production capacity and management services through purchasing bulk water from the NWSDB. For example, the KMC has purchased 0.4 million cubic meters of water from the NWSDB to supply the quantity of water demanded by the KMC consumers.[1]

Table 17.3: Some statistics relevant to water supply at KMC

	2007	2008
Number of connections	26500	27441
Revenue–Billed (Mil.RS.)	261.17	276.35
–Collected (Mil.RS.)	243.83	272.02
Cost of production (Mil.Rs.)	89.88	97.86
Development expenditure (Mil.RS.)	4.42	9.30

Source: Record books of KMC.

Financing and Cost Recovery

Access to Foreign Funds and Financial Situation

To fulfil the MDG targets, the NWSDB has started numerous projects mainly funded by foreign donors, such as the Japan Bank for International Co-operation (JBIC), Asian Development Bank (ADB), United Nations Development Program (UNDP), and the Danish International Development Agency (DANIDA). In 2007, the NWSDB obtained SLR 14,166 million (US$ 111 million)[2] as foreign funds and the government of Sri Lanka provided 4,587 million rupees (US$ 36 million) as counterpart funds (NWSDB 2007). By the end of 2007, the NWSDB was handling 15 foreign-funded projects. Figure 17.2 shows a sharp increase in foreign funds, coming in as grants and loans, especially during the period 2000–07. In 2007, the Board has obtained SLR 36,338 (US$ 284 million) and 12,841 million (US$ 101 million), respectively, as foreign grants and loans respectively (NWSDB 2007). The interest payments by the board have gone up by almost 600 percent over the period of 1995 to 2007 (see Figure 17.3). The level of high interest on loans along with those of foreign grants and loans indicate the NWSDB's dependence on foreign funds.

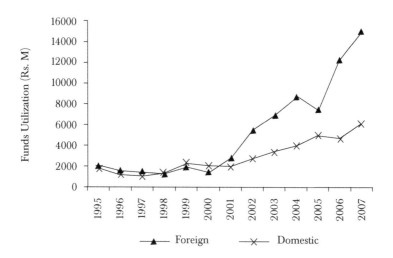

Figure 17.2: Receipt of funds by NWSDB (1995–2007)

Source: Adopted from the NWSDB Annual Reports (various issues from 1987 to 2007).

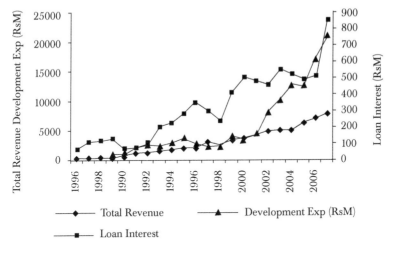

Figure 17.3: Total revenue, development expenditure and loan interest during the period from 1986 to 2006

Source: Adopted from NWSDB Annual Reports (various issues from 1987 to 2007).

Unlike the NWSDB, the KMC does not have access to foreign funds for water supply projects. Instead, the financing is covered mainly through the water revenue, which was 243 and 272 million rupees (US$ 1.91 and 2.14 million), in 2007 and 2008, respectively. The collection was more than 90 percent both years, indicating the KMC's efficiency of bill collection.

Cost Recovery

The revenue for the NWSDB is the lump sum amount collected at the time of water connection and the monthly volumetric charge. The connection charges have been increasing over time. For example, the charges have increased from SLR 14,500 (US$ 114) in 2006 to SLR 20,500 (US$ 161) in 2007. This could prove to be a barrier in achieving the MDGs by excluding the marginalized and poor groups from getting a piped-water connection. Thus, there should be some compensatory measures to avoid the initial burden on poor people in getting access to the drinking water.

In Sri Lanka there are two categories of tariff structures, namely, domestic and non-domestic. The domestic sector has a two-part tariff system while the non-domestic has a fixed rate. At the NWSDB, there have been significant changes in the domestic tariff structure over the years. Until October 1997, income equity among consumers appears to have been an important consideration. However, under the new policy (prevalent during January 1997 to December 2000), revenue raising (and gradual reduction of cross-sector subsidization) seems to have been a dominant criterion in setting water charges (see Table 17.4). Therefore, this policy was likely to have some serious implications for income equity among consumers. Realizing the necessity of equity in water supply, the tariff reform in January 2001 brought back the fixed rate of the first block to a volumetric charge (For more details in tariff reforms, see Hussain et al. 2002). Again in 2009, tariff rates were substantially increased for both domestic and non-domestic sectors. In February 2009, water consumption was categorized into blocks and a service charge (that increases with volume of consumption) was introduced, instead of continuing with the fixed charge. Thus, compared to the tariff rates imposed in 2005, the new tariff structure is expected to increase the revenue substantially.

Similarly, the KMC also collects their revenue through an initial connection charge–a fixed charge depending on the diameter of the water connection[3]–and a volumetric charge for water consumption using increasing block rates (see Table 17.4). The KMC tariff structure is simpler and easier to understand. Unlike the NWSDB, it has no

Table 17.4: Domestic tariff rates of water of NWSDB and KMC during the period 1994 to 2009

Tariff blocks	NWSDB						KMC
	1994	Jan. 1997	Oct. 1997	1999	2005	2009	Current rates
0–10	0.75	0.60	25.00*	30.00*	1.25	5.00	3.00
11–15	1.30	1.50	1.80	2.50	2.50	15.00	8.00
16–20	1.30	1.50	1.80	2.50	8.50	30.00	10.00
21–25	4.80	5.00	6.00	7.50	30.00	50.00	25.00
26–30	4.80	9.60	12.00	15.00	50.00	75.00	30.00
31–40	9.40	12.50	15.00	18.00	60.00	90.00	40.00
41–50	12.00	18.00	20.00	20.00	70.00	105.00	50.00
> 50	25.00	32.50	35.00	35.00	75.00	110.00	60.00

Note: *Consumers were charged a fixed rate regardless of the amount of water consumed.

service charge and its block rates are much lower. For example, KMC consumers will pay only SLR 170 (US$ 1.34) for 20 m^3 of water per month where as NWSDB consumers will have to pay SLR 355 (US$ 2.79) for the same consumption.

The KMC is managed by authorities who are the elected representatives of the voters within the Kandy municipality. Therefore, this political authority has to be sensitive about their vote base and, as a result, has to try and provide good service at a reasonable price. Sometimes, cross-subsidies from other revenue sectors are being used to set-off the losses from water services. Though the KMC appears to balance the day-to-day cash flow in running the water supply services, there are unanswered questions on how it maintains its infrastructure without any investments for maintenance and gradual replacement. In contrast, the compulsion to reduce expenditure per unit of water provided by the NWSDB is less since the affected parties have less control on the officials.

Institutional Collaboration

The KMC is aware that close collaboration in handling water and wastewater management is necessary. For example, a water purification plant was built by the NWSDB at Katugastota with the agreement that the KMC could obtain required water from them; consequently, 10,000m^3 of water is purchased daily at the rate of. SLR 16 (US$ 0.13) per m^3 from there[4]. For the STP, these two organizations have signed an MOU to carry out the work. An awareness program on the STP is also being conducted jointly by these two institutions. According to the MOU, all the construction work will be handled by the NWSDB and the management of the plant (within first two years) will be handled jointly, before the management is fully handed over to the KMC. However, even after the handing over, technical repairs will be done by the NWSDB.

Cost Recovery for a Sustainable WSS Services

The NWSDB has been increasing its efficiency through various means (discussed earlier) and depends upon foreign projects for expansion–which indicates that the institution is financially unstable in the long run. Thus, this study suggests that the NWSDB should develop strategies to reduce expenditure or increase the revenue or a combination of both.

Strategies to Reduce Expenditure

It has been shown that Non-Revenue Water (NRW) has continued to be a significant contributor of the cost of production–as the cost incurred due to NRW is increasing over time (see Table 17.5). By cutting down NRW, the cost of production can be substantially reduced and, consequently, the savings can be used effectively for further development activities in the sector. However, it has been unable to reduce the NRW due to various reasons, such as provision of public stand posts, leakages during the conveyance, and errors in meters. It has also been revealed that a comprehensive assessment is necessary to investigate the critical means of leakages; during a discussion with the KMC it became clear that no assessment was done to evaluate the percentage of water that leaked out at each point during the conveyance from storage to consumer meters. By targeting these important points of water losses, the NRW can be effectively reduced.

The expenditure could be further decreased by reducing the number of new customers served by the NWSDB, thereby reducing the need for capital investments on large scale projects. It is commendable that NWSDB has started several projects using participatory and demand response approaches (DRA) through the mobilization of community resources.

Rainwater harvesting (RWH)–also considered as a cost-cutting practice of WSS–is becoming popular in the rural areas (more prominently with the scarcity of water) as well as the urban areas. Application of RWH in private companies has shown that nearly 50–75 percent of the monthly requirement of non-drinking water can be fulfilled from RWH, reducing the monthly water bill by 30–60 percent (Ariyananda 2004).

Table 17.5: NRW and incurred cost during the period from 2003 to 2007

	2003	2004	2005	2006	2007
NRW (000'm³)	124,442	123,959	96,690	136,630	140,491
Direct cost (million rupees)	1406.5	1648.4	1290.1	2040.4	2377.9
Additional cost born by consumers for every unit of consumption (rupees)	6.04	6.76	6.76	7.82	8.37

Source: NWSDB Annual Report 2007.

One of the most important strategies to reduce the expenditure of WSS could be the use of local technology. At present, the NWSDB carries out new development activities using their own engineers and the materials available in the market. With local technology, the cost of providing services to a customer is substantially low compared to a foreign-funded project where prices include the imported equipments with high costs and elevated estimates, not to mention the exceptionally high salaries of foreign consultants.

Strategies to Increase Revenue

Increasing tariff has been the practice to increase revenue, with or without the consideration of equity over the years (see Hussain et al. 2002). The NWSDB had substantially increased tariff rates in February 2009. Using the data for the months of January to June (in 2009) and an average number of consumers of each block (of these six months), the revenue was calculated (see Table 17.6)– which shows that under the new tariff system, the revenue for the NWSDB has increased by SLR 5,648 million (US$ 44 million). In contrast, if all the block prices were increased by 50 percent there would only be an increase of SLR 1,600 million (US$ 12.5 million) in the revenue. This increment in the revenue is sufficient to cover the budget deficit incurred in 2007.

The NWSDB has formulated its corporate plan for the period 2007–11 (NWSDB Corporate Plan 2007–11). Accordingly, WSS services have to be developed using domestic and foreign funds, which are coming in through projects. By now some of these projects have already begun, some have received Cabinet approval while

Table 17.6: Distribution of revenue among the blocks at different tariff rates

	Revenue at each block (million rupees)									
	0–0	1–10	11–15	16–20	21–25	26–30	31–40	41–50	51–100	> 100	Tot-al
Old Tariff	4	15	15	21	49	52	69	36	56	17	3958
New Tariff	4	21	35	62	122	123	168	93	128	45	9606
50% increase of prices all blocks	4	15	17	26	69	76	103	54	84	20	5562

Source: Authors' calculations based on monthly consumption data at the NWSDB during May 2009 to April 2010.

others are under negotiation. Whatever the stage of the projects, the corporate plan does not have strategies to utilize the augmented local funds generated through increased water tariff. However, it has been mentioned that minor rehabilitation and renovation would be financed through locally-generated funds. There are strategies to implement cost effective operations through increased labour productivity, energy efficiency and treatment process. However, there is no mention about the use of locally-generated funds, as well as of local technology, whilst reducing the dependence on foreign funds. Instead, the NWSDB has planned mostly to depend on foreign funds for future developments; the expected funds in 2011 are in the range of SLR 40,000 million (US$ 313 million).

When consumption data is reviewed over time, there is no significant reduction in consumption due to the price hike. Therefore, one would argue that there is a potential to further increase the tariff rates without reducing the consumption. However, there is an equity issue which needs serious consideration as it is a great financial burden on the poor. For example, a family with a monthly income of SLR 3,000 (US$ 23.5) will have to pay 11 percent of their income for water according to the new tariff rates (SLR 117 [US 4 0.92] per month has been increased to US$ 355 [US$ 3.8]). However, at lower blocks (at least up to 20 m³ per month) consumers are quite inflexible

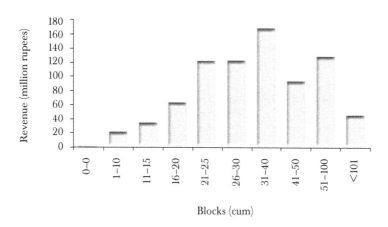

Figure 17.4: Distribution of revenue among blocks

Source: Authors calculation from monthly consumption data at the NWSDB during May 2009 to April 2010.

to the price; but non-reduction in the consumption does not mean that they do not feel the burden of the price hike. The distribution of the revenue over the blocks has shown that 85 percent of the revenue[5] is gained from the consumers who use more than 20 m^3 of water (see Figure 17.4).

Governance

As conceptualized in Figure 17.3, management and financing are considered as internal factors whilst governance is as an external factor affecting the sustainability of the water institutions. The following section tries to explain how the governance issues, as proposed by the 5th WWF, can influence the sustainable urban water supply.

Right to water and access to sanitation

The NWSDB has taken steps to provide facilities and service support to the poor and marginalized communities. Government and donor organizations provide financial support (through the NWSDB) for the community-based water supply and sanitation facilities to rural sections. In collaboration with the Ministry of Health, awareness programs on health and sanitation are also provided for schools and communities. However, in tariff settings for urban water supply it seems that they have neglected the fact that water is a human right despite the subsidy given to *Samurdhi*[6] holders. In contrast, the KMC keeps lower water tariff rates (as compared to the NWSDB) and considers water as a human right. It also provides awareness programs in collaboration with the NWSDB.

Transparency and Accountability

The NWSDB, in particular, is trying to ensure accountability and transparency through various means—as indicated in its corporate plan 2007–11—which includes: (*a*) improving financial control to ensure proper management of funds; (*b*) decentralizing financial and administrative authority in line with the responsibility to form Strategic Business Units ([SBUs] a profit center which focuses on product offering and market segment) in Regional Support Centres (RSCs); (*c*) developing, improving and implementing a comprehensive assets management plan; (*d*) improving the flow of management information and co-ordination among divisions to enable planning and monitoring of activities; and (*e*) meeting the debt service obligation of subsidiary loan agreements with the government. In contrast, the KMC doesn't have any corporate planning.

The administration and management of the NWSDB has undergone progressive transformation, aimed at achieving greater operational efficiency of RSCs through decentralization of operational functions. While the overall revenue collection is managed at the head office, operational functions–including planning and design of development works and financial administration and management– are gradually transferred to the RSC, which independently administers its operation and maintenance budget while billing and metering are further decentralized to district offices. Finally, the NWSDB expects a greater accountability and efficiency at the RSC level (through decentralization) that enables a management and information system capable of accounting expenditure and revenue for each water supply system (NWSDB Annual Report 2007).

Reforms, Regulations and Policies

In spite of the attempts made to ensure accountability, the NWSDB continues to depend on foreign funds for development and increase the revenue by increasing the tariff rates at regular intervals, all the while acting as an autonomous body. However, consumers are not expected to bear the burden of the inefficiencies of institutions. Currently, the annual report is the only document that is produced by the NWSDB, which also has to be defended to the Committee on Public Enterprise (COPE).[7] The consumers have no mechanisms to either intervene or air their grievances. The activities of the NWSDB are not regulated and, as a result, there is a greater need for an independent regulatory authority to ensure accountability. Such a mechanism could also facilitate the participation of stakeholders in the overall governance of the water supply and sanitation sector.

Privatization, PPPs and PUPs

Due to inherent inefficiencies and the pressure from donors, many developing countries privatized their water utilities. However, 80 percent of such private investments were fraught with regulatory failures compounded by problems of information, incentives, and credible commitment (Araral Jr. 2008). Consequently, public– private partnerships (PPPs)–a co-operation between different actors, usually one each from the private and the public sectors, who work together to promote a common objective–became popular. Yet, PPPs also have their own set of problems, such as transaction costs, contract failure, renegotiation, complexities of regulation, commercial opportunism, monopoly pricing, commercial secrecy, currency risk, and lack of public legitimacy (Lobina and Hall 2006). Therefore, public–public participations (PUPs), which are a collaboration of two or more public units, were introduced to overcome the problems associated with PPPs. Furthermore, it was observed that PUP models appear to work well in the Sri Lankan context, especially in the rural water supply sector.

Conclusion

Institutionally, the NWSDB appears to be strong with highly motivated and skilled staff including senior level professionals, who are well educated in various disciplines, and middle-level officers, who are technically qualified with wide experience in water supply and sanitation. The NWSDB is continuing to increase its staff with increasing development activities, though the number of employees per connection has been decreasing–indicating that the organization is expanding its capacity while increasing service efficiency. New interventions to introduce information technology (in core activities) to improve operational efficiency, skill development through training, online payment system, and new infrastructure facilities, among others, have further strengthened the NWSDB.

However, there are issues with regard to its financial sustainability. The total expenditure, including increased cost of production, personal expenditure and increased interest payments of the loans taken, is continuing to deviate from the revenue generated over the years. Continuous dependence on foreign-funded projects will further aggravate this situation unless some corrective measures are taken. Encouraging communities to rely on community-based water supply schemes, adopting water saving techniques (such as rainwater harvesting), commissioning locally funded WSS schemes, and tariff revisions to reduce the gap between expenditure and revenue (while looking after the poorer sections) are important strategies that require serious attention. The governance of such an important national institution should be improved to guarantee reasonable services to its customers. The KMC, on the other hand, has many problems including the lack of funds for infrastructure improvements/rehabilitation along with poor human resources. Therefore, the decision to hand over a sophisticated STP to such an organization needs to be strongly reviewed. However, their legitimate role as a local service provider can not be overlooked and, hence, new institutional arrangements (in collaboration with the NWSDB) need to be explored.

Notes

1. Personal communication with Mr N. G. C. Fernando, Assistant Engineer, NWSDB, Katugastota, Kandy.
2. Converted at the annual average exchange rate of 2012, that is, US\$ = SLR 127.
3. Generally, the (pipe) diameter of a domestic connection is half an inch and the fixed charge is SLR 50 (US\$ 0.39).
4. This also indicated that the KMC is did not have the capacity of producing the water quantity that was demanded.
5. The monthly average revenue from consumers who use over 20 m³ per month is SLR 687 million (US\$ 5 million) from the total of SLR 800 million (US\$ 6 million).
6. A national program for poverty alleviation—where a monthly payment is made by the Government of Sri Lanka to families below the poverty line.
7. The Committee on Public Enterprises was established on June 21, 1979 to ensure the observance of financial discipline in Public Corporations and other semi-governmental bodies in which the government has a financial stake.

References

Asian Development Bank (ADB). 2007. "Sri Lanka Assistance Program Evaluation: Water Supply and Sanitation Sector Assistance Evaluation." Evaluation Working Paper, Operation Evaluation Department, ADB Manila, Philippines.

Araral Jr, E. 2008. "Public Provision for Urban Water: Getting Prices and Governance Right," *Governance*, 21(4): 527–49.

Ariyananda, T. R. 2004. "Rainwater Harvesting for Urban Buildings in Sri Lanka." Paper presented at the Rainwater Harvesting Forum of the 4th IWA World Water Congress, September 20–24, Marrakech, Morocco.

Brown, R. R., M. Mouritz and A. Taylor 2006. "Institutional Capacity," in T. H. F. Wong (ed.), *Australian Runoff Quality: A Guide to Water Sensitive*

Urban Design. Barton, Australian Capital Territory: Institute of Engineers Australia and E A Books.

Daigger, G. T. 2009. "Evolving Urban Water and Residuals Management Paradigms: Water Reclamation and Reuse, Decentralization and Resource Recovery," *Water Environment Research*, 81(8): 809–23.

de Loë, R. C., S. E. Di Giantomasso, and R. D. Kreutzwiser. 2002. "Local Capacity for Groundwater Protection in Ontario," *Environmental Management*, 29(2): 217–33.

de Loë, R. C. and D. K. Lukovich. 2004. "Groundwater Protection on Long Island, New York: A Study in Management Capacity," *Journal of Environmental Planning and Management*, 47(4): 517–39.

de S. Ariyabandu, R. 2006. "Politics of Water: Policies and Institutional Reform." Paper presented at the Second South Asia Research Conference on "Water Supply, Sanitation and Watershed Management in South Asia," September 24–26, organized by the Post Graduate Institute of Agriculture, University of Peradeniya, Kandy, Sri Lanka.

Dharmasena, G. T. 2005. "Inadequacies in Draft Water Policy and Prerequisite: Hydrological Aspects," in *Proceedings of the Consultation on National Water Policy*, June 9, Sri Lanka Foundation Institute, Colombo, Sri Lanka.

Hall, D., Jane Lethbridge and Emanuele Lobina. 2005. "Public–Public Partnerships in Health and Essential Services" Discussion Paper 23, Public Services International Research Unit (PSIRU), University of Greenwich, Available at http://www.equinetafrica.org/bibl/docs/DIS23pub.pdf (accessed May 14, 2013).

Hussain, I., S. Thrikawala and R. Barker. 2002. "Economic Analysis of Residential, Commercial and Industrial Use of Water in Sri Lanka," *Water International*, 27(2): 183–93.

Istanbul Water Guide. 2009. Ministry of Foreign Affairs of Turkey and World Water Council. Available at http://www.unido.org/fileadmin/user_media/Services/Environmental_Management/Water_Management/IstanbulWaterGuide.PDF (accesed May 3, 2013).

Ivey, J. L., R. de Loë, R. Kreutzwiser, and C. Ferreyra. 2006. "An Institutional Perspective on Local Capacity for Source Water Protection," *Geoforum*, 37(6): 944–57.

Lobina, E. and D. Hall 2006. "Public–Public Partnerships as a Catalyst for Capacity Building and Institutional Development: Lessons Learned From Stockholm Vatten's Experience in the Baltic Region." Paper presented at IRC and UNESCO-IHE Symposium on "Sustainable Water Supply and Sanitation: Strengthening Capacity for Local Governance," September 26–28, Delft, The Netherlands. Available at http://www.irc.nl/page/31026 (accessed March 2009)

Narayanan, N. C. and S. Thrikawala. 2011. "Aid, Technology and Project

Dependence: Case of Institutional Weakening of Water Sector from Sri Lanka," *SAWAS*, 2(2): 59–74.

National Water Supply and Drainage Board (NWSDB). 2007. *Annual Report of the National Water Supply and Drainage Board, Sri Lanka.* Available at http://www.waterboard.lk/scripts/ASP/publications.asp (accessed May 3, 2013).

———. 2007–11. *Corporate Plan of the National Water Supply and Drainage Board 2007–11, Sri Lanka.* Available at http://www.waterboard.lk/scripts/ASP/publications.asp (accessed on May 3, 2013).

Ministry of Water Supply and Drainage. 2007–11. *Corporate Plan of the National Water Supply and Drainage Board 2007–11.* Colombo: Ministry of Water Supply and Drainage.

Natural Resources Management Services (NRMS). 2005. "The Environmental Impact Assessment of Kandy City Wastewater Disposal Project," University of Peradeniya, Sri Lanka.

Rogers, P. 2002. *Water Governance in Latin America and Caribbean.* Washington, DC: Inter-American Development Bank.

Saleth, R. M. and A. Dinar. 2004. *The Institutional Economics of Water: Cross-Country Analysis of Institutions and Performance.* Cheltenham, UK: Edward Elgar.

Samad, M. 2005. "Water Institutional Reforms in Sri Lanka," *Water Policy*, 7: 125–40.

Solones, M. and A. Jouravlev. 2006. *Water Governance for Development and Sustainability.* Research Report 111. Santiago, Chile: La Comisión Económica para América Latina (CEPAL).

van de Meene, S. and R. Brown. 2007. "Towards an Institutional Capacity Assessment Framework for Sustainable Urban Water Management," in *Proceedings of the 13th International Rainwater Catchment Systems Conference and 5th International Water Sensitive Urban Design Conference*, August 21–23, Sydney, Australia.

van de Meene, S. 2008. "Institutional Capacity Attributes of Sustainable Urban Water Management: The Case of Sydney, Australia." Paper presented at the 11th International Conference on Urban Drainage, Edinburgh, Scotland, August 31 to September 5. Available at ttp://www.hidro.ufcg.edu.br/twiki/pub/ChuvaNet/13thInternationalConferenceonRainwaterCatchmentSystems/VandeMeene.pdf (accessed June 21, 2010).

Wong, T. H. F. 2006. "Water Sensitive Urban Design: The Story So Far", *Australian Journal of Water Resources*, 10(3): 213–21.

About the Editors

Vishal Narain is Associate Professor, School of Public Policy and Governance, Management Development Institute (MDI), Gurgaon, India. He holds a PhD from Wageningen University, The Netherlands. His academic interests are in the interdisciplinary analyses of public policy processes and institutions, water governance and peri-urban issues. He has published chapters in edited volumes and several articles in leading journals, such as *Water Policy, Water International, Environment and Urbanization* and *South Asian Water Studies*. He received the S. R. Sen Prize for the Best Book on Agricultural Economics and Rural Development (2002–3) conferred by the Indian Society for Agricultural Economics for his book *Institutions, Technology and Water Control: Water Users' Associations and Irrigation Management Reform in Two Large-scale Systems in India* (2003). He was a lead author for a chapter on human vulnerability to environmental change for *Global Environment Outlook (GEO)–4*, the flagship publication of the United Nations Environment Programme.

Chanda Gurung Goodrich is Principal Scientist, Empower Women, International Crop Research Institute for the Semi-Arid Tropics (ICRISAT), Hyderabad, India. She holds a PhD from the School of International Relations, Jawaharlal Nehru University, New Delhi, India. Her interest is in gender research in natural resource management, agriculture and livelihoods, especially for small holder farmers. She has extensive experience in gender and participatory research and development, having worked as a researcher and consultant in the not-for-profit sector with various international and regional organizations in South Asia, specializing in integrating social and gender equity into development programs and projects. She also has published several papers in journals, and chapters in edited volumes on gender and agriculture/natural resource management.

Jayati Chourey is Assistant Professor, Department of Energy and Environment, Symbiosis Institute of International Business (SIIB), Symbiosis International University (SIU), Pune, India. Previously, she worked with South Asia Consortium for Interdisciplinary Water Resources Studies (SaciWATERs), Hyderabad, India, as a Senior Fellow (Education and Networking) and was also responsible for managing SaciWATERs-CapNet Network (SCaN). She holds a PhD in Ecosystem-based Water Resources Management from the Indian Institute of Forest Management, Bhopal. The focus areas of her work include natural resource management, environmental governance, water, environmental monitoring, energy, health, and livelihoods. She was an Environment Equity & Justice Partnership (EEJP) Fellow (2005–6), a program supported by the Ford Foundation. She is also associated with various environmental and social development forums.

Anjal Prakash is Executive Director, SaciWATERs, Hyderabad, India. He is also Project Director of "Water Security in Peri-Urban South Asia," a project funded by the International Development Research Centre (IDRC). He holds Master's degree from Tata Institute and Social Sciences (TISS), Mumbai, India, and PhD in Social and Environmental Sciences from Wageningen University, The Netherlands. He has worked extensively on the issues of groundwater management, gender, natural resource management, and water supply and sanitation, and currently focuses on policy research, advocacy, capacity-building, knowledge development, networking and implementation of large-scale environmental development projects. He is also the author of *The Dark Zone: Groundwater Irrigation, Politics and Social Power in North Gujarat* (2005). His recent edited books include *Interlacing Water and Human Health: Case Studies from South Asia* (2011) and *Water Resources Policies in South Asia* (2013).

Notes on Contributors

Shrinivas Badiger is Fellow, Center for Environment and Development, Ashoka Trust for Research in Ecology and the Environment, Bangalore, India. He has a PhD in Soil and Water Resources Engineering from University of Illinois, USA. His research activities anchored in the water sector use systems approach linking biophysical and socio-economic processes to understand the context of resource depletion within a larger framework of sustainable environments and human well-being. His current work focuses on understanding the impact of climate change and variability, including that of extreme rainfall events on hydrologic flow regimes, flood regulation and adaptation in irrigated agriculture.

Gongsar Karma Chhopel is currently Head, Water Resources Co-ordination Division, National Environment Commission, Thimphu, Bhutan. He completed his BS degree in Limnology from Southwest Texas State University, USA, and MS degree in Aquatic Biology from Texas State University, USA. He is He is the Country Focal Person on all issues related to water and its development. He is also Executive Member to the Bhutan Water Partnership; National Focal Point to the Asia Pacific Network for Global Change Research; Chairman of the Technical Advisory Panel; Team Leader on Water Thematic Group for the Bhutan Climate Summit 2011; and global contact point for issues on water in the region.

Prakash Gaudel is Environment Specialist, Nepal Electricity Authority, an undertaking of the Government of Nepal. He completed his Master's degree in Interdisciplinary Water Resources Management from Nepal Engineering College under the South Asian Water (SAWA) Fellowship of SaciWATERs, Hyderabad, India. He also holds a Master's degree in Environment Science (2009) from Kurukshetra University, India, under the Silver Scholarship

Scheme of Embassy of India in Nepal. His research interests include issues related to the water supply sector, transboundary water management, climate change and adaptation, and environmental impact assessments.

Smitha Gopalakrishnan is Assistant Professor, National Institute of Technology, Calicut, India. She has a Bachelor's degree in Architecture from the same institute and a Master's degree in Urban and Regional Planning from CEPT University, Ahmedabad, India. Her Master's thesis dealt with institutional restructuring for improved urban water supply service delivery. She also works in the areas of urban food security, urban services (water supply) delivery, infrastructure provisioning, and environmental policy-making.

L. H. P. Gunaratne is Professor, Department of Agricultural Economics and Business Management, Faculty of Agriculture, University of Peradeniya, Kandy, Sri Lanka. Currently, he serves as Director, Agribusiness Centre, University of Peradeniya, and Secretary, Board of Study in Agricultural Economics, Postgraduate Institute of Agriculture. Over the last 16 years, he has worked on Environmental and Resource Economics through his teaching, research and outreach activities. He has authored several articles in refereed journals, book chapters, conference proceedings and abstracts, and books.

E. R. N. Gunawardena is Senior Professor, Agricultural Engineering, University of Peradeniya, Kandy, Sri Lanka. He graduated in Agriculture from the same university and obtained his Master's degree and PhD in Soil and Water Engineering from Cranfield University, UK. In his long career of 30 years, he was also the Head of the Department of Agricultural Engineering, Dean of the Faculty of Agriculture, and Deputy Vice-Chancellor of the University of Peradeniya. He has published extensively in the areas of hydrological simulation, irrigation, drainage, watershed management and contributed to the formulation of protected areas of Sri Lanka and the forestry sector master plan. His current research interests are in the area of Integrated Water Resources Management (IWRM). He was the first country co-ordinator of the CapNet, a UNDP project on capacity-building in IWRM and later became the Executive Director of SaciWATERs, Hyderabad,

India, and the Project Director of the "Crossing Boundaries Project," a regional capacity-building project on IWRM, Gender and Water.

Hamidul Huq is Founder and Chairman, Institute of Livelihoods Studies (ILS) and Founder, Unnayan Shahojogy Team (UST), a national NGO in Bangladesh. Currently, he is engaged in interdisciplinary research and policy advocacy in the fields of water, ecosystem services, climate change and livelihoods in coastal zones. He also teaches Disaster Management at BRAC University as part-time Professor. He has a PhD in Rural Development Sociology from Wageningen University and completed his Master's degree in Social Welfare from the University of Dhaka. Dr Huq has been actively involved in research and policy advocacy and has worked on the Integrated Coastal Zone Management Plan project of Bangladesh (2002–6), which produced the Coastal Zone Policy, Coastal Development Strategy, and a database of the coastal zones of Bangladesh.

Deepa Joshi is Assistant Professor, Conjunctive Water Management and Conflict, South Asia, Irrigation and Water Engineering Group, Center for Water and Climate, Wageningen University, The Netherlands. She has a PhD from the University of Southampton, UK, and over 10 years of program management and policy research experience in gender and natural resource management governance. She has also worked extensively in South Asia, Africa and Latin America.

Seema Kulkarni is founding member and Senior Research Fellow of the Society for Promoting Participative Eco-system Management (SOPPECOM), Pune, India. She has been working in the area of natural resource management and rural livelihoods since 1991, and is actively engaged in policy advocacy and gender training in the area of gender and rural livelihoods. She has been consistently associated with the women's movement in Maharashtra, particularly in south Maharashtra on the question of deserted women as part of the Stree Mukti Sangharsh Chalwal or the Women's Liberation. She has published several articles on the issues of gender, water and rural livelihoods, and contributed chapters on the same themes in edited volumes.

M. Shahjahan Mondal is Professor, Institute of Water and Flood Management (IWFM), Bangladesh University of Engineering and Technology (BUET), Dhaka. He did his undergraduate and postgraduate studies from the same university and obtained his PhD from Central Queensland University, Australia. Dr Mondal attended training programs on water governance, water and equity, and water and rights, and has published several articles on water resources planning and management in leading journals including those of ASCE, Elsevier Science, Springer and Wiley & Sons. His current research interest is in the institutional aspects of water and related resource management.

Dhruba Raj Pant is currently associated with Jalsrot Vikas Sanstha (JVS)/Global Water Partnership (GWP)/Nepal. He has a PhD in Development Sociology from Wageningen Agricultural University, The Netherlands. He has been involved in studies on natural resource management and associated institutions for poverty alleviation through improvement in livelihood opportunities for various communities mainly focusing on IWRM and climate change effect in a river basin. He has also been engaged in the formulation of policy for natural resources management along with the design and implementation of water resource management programs in multidisciplinary teams.

R. Parthasarathy is Professor, Faculty of Planning and Public Policy, CEPT University, Ahmedabad, India. He has about 25 years of teaching and research experience and has been a Visiting Scholar at the University of California, Berkeley; Professor at and later Director of Gujarat Institute of Development Research, Ahmedabad. He has been a member of many government committees and also a consultant to DFID, ADB and other national and international organizations. He also works closely with NGOs. His research interests include water management, large-scale irrigation systems, natural resources and environmental economics. He has published three books and over 70 articles.

Iswaragouda Patil is Senior Research Associate, Center for Environment and Development, Ashoka Trust for Research in Ecology and the Environment, Bangalore, India. He has a Master's degree in Economics, and has more than 12 years of field experience working on issues of natural resources management and livelihoods. His work has been focused on understanding the impact of land-use change on watershed services and specifically on livelihoods of downstream communities. His recent work explores the trade-offs in access to ecosystem services to various stakeholders in complex forest and agro-ecosystems, including impact of climate change on agriculture.

Soumini Raja is Assistant Professor, College of Architecture, Trivandrum, and PhD Fellow, CEPT University, Ahmedabad, India. She completed her Bachelor's degree in Architecture from College of Engineering, Trivandrum, and Master's degree in Urban and Regional Planning from CEPT University, Ahmedabad. Her work and research has been in the field of climate change and coastal zones. She has also presented and published papers at various national and international conferences.

Divya Badami Rao has been associated with organizations working at the interface of environment and development following her Master's degree in Social Work from Tata Institute of Social Studies (TISS), Mumbai, India, where she specialized in urban and rural community development. She has previously worked on issues of environmental governance and regulation through campaign-based advocacy and research at Kalpavriksh Environmental Action Group, New Delhi. Her most recent work is a project on evaluating the missions under India's National Action Plan for Climate Change, jointly undertaken by the Center for Development Finance (CDF), Institute for Finance Management and Research (IFMR) and Indian Institute of Technology (IIT), Chennai.

Mashfiqus Salehin is Professor, Institute of Water and Flood Management (IWFM), Bangladesh University of Engineering and Technology (BUET), Dhaka. He completed his Bachelor's degree in Civil Engineering and Master's degree in Water Resources Engineering from BUET and his PhD from Northwestern University, USA. He teaches postgraduate courses on Interdisciplinary Field Research Methodology, Hydrogeology and Groundwater, and Groundwater Resource Assessment. Dr Salehin has conducted research on a variety of issues including modeling of water resources systems (at regional and national levels), stream–groundwater interactions, and mechanisms of water-related natural disasters. His research interest lies in interdisciplinary approaches to water management. He is also the Co-ordinator (from BUET) of the "Crossing Boundaries Project" of SaciWATERs, Hyderabad, India. Additionally, he has published several research reports and scientific papers in national and international journals.

Khem Raj Sharma is Director, Center for Postgraduate Studies, Nepal Engineering College, Kathmandu. He has specialized in land and water resources engineering and worked for the Department of Irrigation (DoI), Government of Nepal, for several years as Research Director and Project Co-ordinator of irrigation development projects. He has also worked as an engineer-in-charge of the Integrated Hill Development Project and the Hill Food Production Project. He has authored two books, edited seminar proceedings, and presented numerous scientific papers in national and international forums. He was Chief Editor of *Irrigation Newsletter* and has taught and guided engineering students on their thesis work.

Bijaya K. Shrestha is President, Settlement–Society– Sustainability (S3) Alliance, a development forum for habitat (a non-government organization). He completed his PhD in Urban Engineering, University of Tokyo, Japan. Previously, he worked with the Ministry of Housing and Physical Planning, Government of Nepal; United Nations Centre for Regional Development (UNCRD), Japan; and various architectural schools in the Kathmandu Valley. He has published chapters in edited volumes, as well as articles in international journals, conference proceedings, local magazines, journals and newspapers. His field of expertise includes sustainable urban development, disaster management, housing, local government capacity-building and development control.

Dibesh Shrestha is Research Associate, Nepal Development Research Institute (NDRI), and is working on a project, "Disaster Risk Reduction and Climate Change Adaptation in Koshi River Basin, Nepal." He has a Bachelors' degree in Civil Engineering and a Master's degree in Interdisciplinary Water Resources Management (IWRM) from Nepal Engineering College, Kathmandu, under the South Asian Water (SAWA) Fellowship of SaciWATERs, Hyderabad, India. He was also involved with the

Kathmandu Valley Water Supply Management Board (KVWSMB) in monitoring of Low Income Consumers' Support Unit (LICSU) program in various part of the Valley and partly engaged in monitoring of licensed tube wells.

Hari Krishna Shrestha completed his Bacherlor's degree in Hydrology from Tarleton State University, USA; Master's degree in Hydrology from New Mexico Institute of Mining and Technology, USA; and PhD from Ehime University, Japan. As a Staff Hydrologist at the Geoscience Consultants Limited in Albuquerque, New Mexico, USA, he developed transient 3D computer model of the groundwater and solute transport of an aquifer at the White Sands Testing Facility of NASA in Las Cruces, New Mexico. Previously, he was Assistant Professor at Nepal Engineering College. After completing his PhD, he was involved in disaster risk reduction activities in Nepal, including public awareness enhancement. As Director, Center for Disaster Risk Studies, he has conducted numerous professional and community level trainings on DRR, focusing on water-induced disasters. He has published more than 36 research and technical papers in national and international journals.

Sushmita Shrestha is Lecturer, Department of Architecture, Khwopa Engineering College, Bhaktapur, Nepal. She got her Bachelor's degree in Architecture and Master's degree in Urban Design and Conservation degrees from the same university. She has also been working as a researcher at Settlement–Society–Sustainability (S3) Alliance. She has already authored dozens of articles, published in edited books, international journals, conference proceedings, local magazines, journals and newspapers. Her field of expertise includes architectural and urban development, urban design, community planning, conservation, and human settlement.

Ashutosh Shukla completed his Master's degree in Land and Water Resources Engineering from University of Philippines in 1989. He joined Nepal Engineering College (NEC) after more than 20 years of teaching and research engagement in Tribhuvan University, Kathmandu, Nepal. His research interests include farmer-led integrated land and water management. He has been working on the systems approach to water resources planning. He has also been responsible as a research co-ordinator in the Interdisciplinary Water Resources Management Program, underway at NEC, in support of the "Crossing Boundaries Project" of SaciWATERS, Hyderabad, India.

Sunil Thrikawala is Senior Lecturer and Head of the Department, Agricultural and Plantation Engineering, Faculty of Engineering Technology of the Open University, Sri Lanka. He also serves as a member of the teaching panel of the Postgraduate Institute of Agriculture and as a Resource Economist of the Center for Environmental Studies, University of Peradeniya, Sri Lanka. He holds BSc and MPhil degrees in Agriculture, MSc in Agricultural Economics and PhD in IWRM. He has published several journal articles and chapters in edited books in the areas of water resource and agricultural economics.

Subodh Wagle is Professor, School of Habitat Studies, Tata Institute of Social Sciences (TISS), Mumbai, and Adjunct Professor, Indian Institute of Technology (IIT), Mumbai, India. He is also Trustee and President of PRAYAS, an independent public-interest policy research and advocacy organization in Pune, India. He has a B. Tech. (Mechanical Engineering) from IIT (Mumbai) and PhD (Energy and Environmental Policy) from the College of Urban Affairs and Public Policy, University of Delaware, USA. He has designed and teaches post-graduate courses in the areas of public policy, infrastructure sectors policy, evidence-based policy advocacy, sustainable development, urban water policy. He has published widely in the areas of water policy, reform, and regulation, electricity reform and regulation, Enron controversy and environmental policy. He has also conducted, and participated in many training programs, workshops, seminars, and conferences.

Sachin Warghade is Assistant Professor, Tata Institute of Social Sciences (TISS), Mumbai and is involved in launching pioneering Masters Programs in Regulatory Governance and Water Policy. He has worked in the field of agriculture, disaster management, water regulation and water policy during his decade-long association with PRAYAS, Pune. Apart from numerous research and publications, his work was instrumental in public-interest protection in regulatory issues of water allocations and tariff. He was a member of the

Planning Commission's XIIth Plan sub-group for development of "Model Bill for State Water Regulatory System."

N. I. Wickremasinghe is Senior Sociologist, Project Management Unit (PMU), Regional Support Center (Central), Kandy, Sri Lanka. Previously, he was engaged with the National Water Supply and Drainage Board, Sri Lanka (Prime National Authority for drinking water supply) with 18 years' experience in project planning, implementation, monitoring and evaluation, public awareness, capacity-building, and training. He has Bachelor's and Master's degrees in Sociology from the University of Peradiniya, and a Post-Graduate Diploma in Community Development from Colombo University. He is engaged in various part-time consultancies and assignments for rural water supply and sanitation improvement programs, developing village development plans, micro-financing, livelihood development, social mobilization, and community training. He has contributed several articles in various journals and books in the area of community involvement of rural water supply sanitation activities.

Index

civil society 2, 4–5, 7, 9, 11, 87, 97, 169, 180, 186, 276, 288–89
Committee on Public Enterprise (COPE) 306, 308n7
Community-based Integrated Natural Resource Management (CBINRM), Nepal: integrated approach in resource management 104; requirements for water resources planning 104; stakeholders in resource management 104; *see also* action research on water resources planning, Nepal
community-based organization (CBO) 57, 60–62, 68–69, 93, 98–99, 115
community empowerment process, in water supply 92–94
Community Water Supply and Sanitation Project (CWSSP), Sri Lanka 54, 55, 92
Convention on Elimination of Discrimination Against Women (CEDAW) 79
Convention on the Rights of the Child (CRC) 79
"crossing boundaries" project ixx
Cullet, Phillipe 36, 42–44

Dalits 8, 23, 26–28, 31, 36–37, 41–49, 114, 128; access to water sources by 39–40; drinking water supply to 42; water policy, impact of 44
Demand Responsive Approach (DRA) 43, 44, 61
Department of Rural Development, Government of Uttar Pradesh 44
Dhaka city, study of governance of rivers: accountability issues 286–87; alarming condition of the rivers 281–82; encroachment of rivers 279–80; institutional capacity and resources, status of

287–88; methodology 276–77; new institutional framework 288–89; present institutional set-up for governance of rivers 283–88; river system in and around the city 274, 277–78; role of multiple stakeholders in water resource management 287; water flow in rivers 280–81; water pollution 278–79
domestic water supply, neoliberal approaches to 43–44; Swajal Project (1996–2002) 44–48
drinking water policy, in India 42
drinking water projects 27, 30, 48, 249
drinking water sector, gender asymmetries in 41–43
Dublin Principles 85
dug wells 57, 60, 65, 66, 137, 139–40, 142–48, 152–55, 256–58, 261, 264, 266

Earthquake Reconstruction and Rehabilitation (ERR) project 27, 33n4
exclusion, politics of 28

Farmer Managed Irrigation Systems (FMIS) 107
Farmer Organizations (FOs) 86, 90, 95, 98
Fawcett, Ben 37, 41
feminist discourses 24
Field Canal Groups (FCGs) 90
food availability 29
foreign investment 4
fresh water 54, 74, 85

Galoya Development program 90
Gelinas, J. B. 1
gender asymmetries, in India's drinking water sector 41–42
gender-caste-water inequities 44

For Product Safety Concerns and Information please contact our EU
representative GPSR@taylorandfrancis.com
Taylor & Francis Verlag GmbH, Kaufingerstraße 24, 80331 München, Germany